KT-559-073

Therapy Pets

Photography by

Donald W. Smith

Therapy Pets

The Animal-Human Healing Partnership

Jacqueline J. Crawford & Karen A. Pomerinke

59 John Glenn Drive
Amherst, New York 14228-2197

Published 2003 by Prometheus Books

Therapy Pets: The Animal-Human Healing Partnership. Copyright © 2003 by Jacqueline J. Crawford and Karen A. Pomerinke. Photographs copyright © 2003 by Donald W. Smith. All rights reserved. No part of this publication may be reproduced, stored in a retrieval system, or transmitted in any form or by any means, digital, electronic, mechanical, photocopying, recording, or otherwise, or conveyed via the Internet or a Web site without prior written permission of the publisher, except in the case of brief quotations embodied in critical articles and reviews.

Inquiries should be addressed to
Prometheus Books
59 John Glenn Drive
Amherst, New York 14228–2197

716–691–0133 (x207). FAX: 716–564–2711.
WWW.PROMETHEUSBOOKS.COM

07 06 05 04 03 5 4 3 2 1

Library of Congress Cataloging-in-Publication Data

Crawford, Jacqueline J., 1952–
Therapy pets : the animal-human healing partnership / by Jacqueline J. Crawford and Karen A. Pomerinke ; photography by Donald W. Smith.
p. cm.
ISBN 1–59102–071–9 (pbk. : alk. paper)
1. Pets—Therapeutic use. 2. Animals—Therapeutic use. 3. Human-animal relationships. I. Pomerinke, Karen A. II. Title.

RM931.A65.C73 2003
615.8'515—dc21 2003005799

Printed in the United States of America on acid-free paper

Contents

Preface

Jacqueline J. Crawford
and Karen A. Pomerinke

The stories in this book, although often astounding, are by no means unique. For every story included there are at least six or seven stories, just as astounding, that did not get told. The amazing healing is usually witnessed by the animal owners and elicited by their certified dog, horse, cat, or llama (among others, one story tells of a certified ferret and a certified pot-bellied pig). Many times the animal handlers did absolutely nothing to elicit the response. If there is some explanation for why animals repeatedly open doors for our physical and emotional healing, perhaps it is that they have a sixth sense that tells them when someone is stressed and in need of their calmness. Certainly not all animals reach out to people. I can only guess that some animals that have trained for their jobs have also been assured that they are safe and loved; it is these animals that are willing to reach out and offer healing to people on their therapy rounds.

There were many Animal-Assisted Therapy (AAT) stories sent in that we could not include. In some cases the need to maintain someone's confidentiality was the impediment. In others, however, it had more to do with my own insistence that we include only stories with real names and photos. I believe that we owe this "realness" to our readers. Still, there have been more times than I want to say that I have had to tell someone we could not use their story, knowing that the story was exactly what we had hoped to include. I hope that there will be a future volume of *Therapy Pets*, so that more people who have experienced the benefits of AAT can tell their stories.

In the long-term interest of expanding the field of AAT, it is absolutely essential that animals be well trained and trustworthy, and that other people can be assured of these skills and traits. National certification is, in my opinion, one way to accomplish this. Also, because the field is still relatively new, people are confused when they see an animal in a therapeutic setting where they are not usually allowed. They are equally confused by the terms used in association with these therapy animals. We have added an appendix (Terms Associated with Therapy Pets) to try to sort out some of the confusion. People assume that if an animal is in a normally animal-free environment, it is a service animal (an animal trained to work with a particular person who has a disability). Even if the AAT handler explains that the animal accompanying them is certified by a national AAT organization, most people misunderstand, forget, or don't listen because they assume they already know what the animal does. We hope this book will clear up some of this confusion.

Credit for these stories goes not only to the authors whose names appear on the cover, but also to those whose beautiful words were a source of inspiration for me, like Josiah's: "The grief that blanketed Manhattan during our time there was so much heavier that the weight of the towers themselves." His musical sense flowed around the story he sent. His photos of Ground Zero are also quite illustrative of AAT, and sometimes haunting.

Credit is deserved for all those who were willing to help this book come into being. They are not only those whose names appear at the tops of the stories and beside the photographs, but others whose names lie beneath the stories. We are thinking of, for example, Jeff Frank of the *North County Times* who, by printing in his newspaper column my request for news of Hannah Redford, was instumental in reuniting Shannon Cote and her daughter, Hannah Redford, with Lee Gaffney and her husband, Tom. If Jeff had not been willing to print this query, Shannon, Hannah, Lee, and Tom would still be no more than fond memories for each other.

We are thinking, too, of Steven L. Mitchell, our editor, whose sense of this book's potential led him to move forward with its publication. Thank you Tom Frank, thank you Steven, and thank you everyone who wrote and called and continued writing, calling, and sending us things each time we requested it. We are deeply and forever indebted to you all, for without you this book would not have been.

This book shows stories and photos of people whose lives were improved by therapy pets, and we hope it is just the beginning. So we ask that if you have a story to share, please send it in! Here is what we need:

- The story
- Permission of the people in the story to publish it
- The opportunity to take photos of the therapy pet with the people in the story.

Please contact us at Jackie Crawford, P.O. Box 10265, Fargo, ND 58106, or e-mail me at jackie@therapypets.com.

Thank you for your help.

Preface

Donald W. Smith

Sometimes the greater part of genius is simply recognizing what is right in front of us. I have learned that the kingdom of animals is one of the great nations of earth, and I have seen how animals are often the center of remarkable healing stories witnessed by human beings.

I think of a little girl, Hannah Redford, who lay in a coma for ten days in New Orleans, Louisiana, being awakened by a little white dog named Molly, that had been rescued from the needle of death in a dog pound. The best efforts of the hospital physicians and nurses, with their many years of medical training from some of the finest universities in the world and with many more years of life-and-death practical experience to guide them, could not bring Hannah out of her coma. On Hannah's day of awakening, it was not medical science that showed its power, but rather the touch of love in the person of Lee Gaffney and her little white mutt, Molly. Standing at the door to Hannah's hospital room, Lee was told by a hospital nurse not to bother coming inside, because "there is little hope in here." Inside, Hannah's mother, Shannon Corte, continued the long bedside vigil of a loving mother's devotion to her sleeping daughter. Upon hearing the conversation at the door, Shannon immediately rose to welcome Lee and Molly, and invited them to visit Hannah. With the simple greeting of dog kisses liberally applied to sleeping Hannah's face, the spark of consciousness began to return and Hannah was awakened by the natural instincts of Molly the therapy dog.

I am mindful also of the story of Amy Roth and her horse, Magic. I was privileged to photograph them at the Clare-B Training Center near Brainerd, Minnesota, home of the Mounted Eagles Therapeutic Riding Program. Therapeutic horseback riding is a perfect example of how genius is often an intangible "something" that is right in front of us just waiting to be recognized. I have learned that riding horses can be the most therapeutic of all possible physical therapies for many persons who suffer from diseases of the muscles and joints. It turns out that the natural, rhythmic movement of horse and rider is now recognized to be more effective and more beneficial as a muscle and joint therapy than many other therapies devised by medical science.

The newly discovered benefits of therapeutic horseback riding represent a perfect, down-to-earth example of something truly marvelous in our observable world.

I am so grateful to Jackie Crawford for researching the subject of animal-assisted therapy, and for presenting this therapy model as a viable and effective potential partner to the traditional treatment of physical and emotional illness. This small collection of animal-assisted therapy stories presented in Jackie's book bears startling witness to the previously unlooked-for benefits and unrecognized potential of animal-assisted therapy in the treatment of human illness. If you are reading these words, chances are good that you, too, have a wonderful story to tell. Perhaps someday you will contact Jackie and share your story with her, and perhaps it may be told in one of her next books.

Hospital administrators, physicians, nurses, and other policymakers have the opportunity to be regarded as insightful and visionary for simply recognizing what is right in front of them—that the marvelous healing potential of animal-assisted therapy can be applied everyday in hospitals and care facilities for the benefit of patients and healers alike. The potential of animal-assisted therapy is a story of genius just waiting to be recognized.

Chapter One

Ben

She knew that Dale had been unlikely to take his own pains seriously. He regularly put up with "the buggers," as he called them. His chief concern was the cows that could not wait for tending until he felt better. Her husband was, she reflected, pretty typical of the farmers she had known all her life, ignoring as best he could the cuts and scrapes, colds and flu, and squeezing pressure across his chest, pressure that several times a week took away his breath and buckled his knees, and which he had done his best to protect her from. Sauntering into her kitchen and joking about being outstanding in his field, he had never hinted at his heart failings. If she had known, she might still have him with her.

Elizabeth Clark had instantly forgiven her daughter's fiancé for not telling anyone. It was just like Dale to swear him to secrecy. Dale, gritting his teeth, had tersely muttered that these durned sweat-breaking chest-squeezings hadn't stopped him before and they weren't about to stop him now. He was not to say a word to anyone, Dale ordered, certainly not to Jennie. It didn't matter that his daughters were now grown; he protected them as closely as the day they were born, as he did his wife of more than thirty years. Jennie's fiancé hesitantly kept his promise until, one day in February, he saw Dale drop like a sack of grain.

After the 911 call and the ambulance ride to Southwest Washington Hospital, Dale Clark was diagnosed with a massive coronary that had permanently and severely damaged his heart. It was not his first heart attack, although it was the first to be diagnosed, and he would endure many more. No more than two weeks of relative peace interspersed each attack. Each time Dale was admitted to the hospital he slid closer to death. Remarkably, these frequent hospital stays also birthed a deep and abiding friendship with a massive, 165-pound brown Newfoundland named Ben.

Ben and his owner, Pat Dowell, had been visiting Southwest Washington Hospital as a certified therapy animal team for about three years when they first met Dale Clark. For reasons Pat still does not understand, Ben's special ability to discern who needed his devoted attention and to remember that person was doubly evident when it came to Dale. When Dale was there, Ben bee-lined for his room without being told where to go. Each time Dale returned to the hospital Ben was there, gazing intently at him as he sat by Dale's hospital bed or, his long, loping tail moving in encouragement, watching Dale fight through his physical therapy. Ben's presence gave Dale the strength to muster his courage and keep trying. It was Ben's photo that Dale insisted sit next to those of his two granddaughters in the hospital room. Any visitor who looked inside the room might see Ben gently place his front paws on the hospital bed beside Dale, his tail slowly waving back and forth. They would then see Dale smile, his eyes light up, and his whole body relax as he stroked Ben's head, massaged his ears, and scratched Ben's neck and chest. During those visits Ben was clearly Dale's dog.

As the months passed and the regular admittances continued, the heart damage became so extreme that doctors were sure Dale was not going to live. Each time they prepared his family for Dale's passing, going as far as removing all life support and making Dale

Dale Clark's granddaughter Adreanne Pepper and Ben.
(Photo by Pat Dowell)

comfortable. Each time, Dale managed to struggle back to some semblance of normal life.

One summer's day, while driving on the freeway, Dale suffered another heart attack and had a terrible auto accident. At the hospital the attending doctor met with Jennie in the hallway. "Do not," he counseled, "let your father know how bad off he is. He wants to live for you, but he is suffering terribly. Let him know he is okay. Let him go." Dale's family agreed to stop all extraordinary measures to revive him. The torturous decision made, they watched as he lay still and colorless, apparently as ready to die as the doctor had indicated. Still, Dale's two daughters were desperate for a way to comfort him, to take away whatever pain he might still be feeling. Remembering that this day was Thursday, the day that Ben came to the hospital, the two conferred briefly and then hurried to the hospital lobby where Ben typically went after his visits to the pediatric and inpatient rehabilitation units.

As Pat entered the lobby with Ben, the young women rushed toward her. Their father was Mr. Clark, they explained, who was in

From left: Pat Dowell, Ben, and Adreanne Pepper. (Photo by Jennifer Pepper)

the Intensive Care Unit and not expected to live. They begged Pat to visit him, believing that Ben's presence would greatly lift the old man's spirits. Pat knew that the ICU was not an authorized area for the animal-assisted therapy program and so regretfully declined. Momentarily discouraged, but not deterred, the women turned to Larry Newell, the hospital security manager who was known to them and who, standing nearby, had overheard their insistent pleas.

In my conversations with Larry about Ben's work at the hospital, he reflected on the first time he saw Ben there. "I walked out of the elevator and saw the hindquarters of this big bear of a dog." As it was his job to ensure hospital safety, Larry naturally investigated the situation. Among other things, Larry learned that Ben was a certified therapy dog working in the hospital. Larry's work raising and training yellow Labs to be bird dogs helped him understand how a dog's intuitiveness could help people in pain. "Hospitals are sometimes a little grim, can be a little depressing," said Larry. "[Ben] seems to sense pain and suffering and goes to that person." As his friendship with Pat and Ben grew, Larry learned that Ben liked

Elizabeth and Dale Clark. (Family photograph)

riding around the grounds in the hospital golf cart and that Ben knew how to give a "hospital bark"—a soft version of his normal booming bark—when he wanted attention. But the ultimate compliment that Larry gave Ben told how much the "big guy" had grown in his estimation: "That dog is more human than animal."

Larry also had become friends with the Clark family over the months that Dale had been in and out of the hospital. He knew that the girls were not exaggerating their father's circumstances and that he needed to do everything he could to help. After some quick phone calls, Larry triumphantly announced that the hospital staff had arranged for Pat and Ben to be allowed into one of the rooms adjacent to the ICU. The trio of women and Ben lost no time moving upstairs to see Dale.

When Pat and her bear of a dog walked into the ICU, Big Ben was enthusiastically greeted by the nurses, aides, and doctors who work there. They gave Ben a round of smiles and hugs as he headed toward Dale's room. While heartening, the welcome that the staff gave Ben paled in comparison to the reception he received from the old man.

As Ben sauntered toward Dale's room, Jennie whispered softly to her father, "Dad, Ben is here." Dale opened his eyes. His face changed from pale and impassive to warm and delighted; he smiled weakly as he saw Ben coming. Dale reached out his arm for Ben. Ben, knowing his part, automatically put his paws up on the bed where Dale could reach him. Within minutes of Ben's arrival, to the amazement of the growing assemblage, Dale was sitting up in bed and talking to Ben.

Here was a man in a great deal of pain, facing certain death, who was suddenly sitting upright in bed, hugging and petting the huge brown Newfoundland with the calm, direct gaze. Dale's family watched the reception with tears in their eyes. Neither were the hospital staff immune to the deep emotional bond they witnessed. Their low murmurs echoed softly down the hall. In the days following, Dale made, yet again, another unexpected recovery.

Across the weeks and repeated admittances, Pat and Ben were there. They spent time with Dale and his family and grew close to them. The family, in turn, shared with Pat how Ben had become a real high spot in Dale's life. Dale, as witness to his honor and devotion to others, had a large and devoted support system. He told every one of them about his giant friend, Ben the dog. It seemed that Pat and Ben had been adopted; they were honored to be a part of the Clark family.

It was on a routine hospital visit one day in October, some eight months after their first visit, that Pat heard Dale had finally succumbed to his amassing heart damage. Because of the special attachment Ben had, not only to Dale but to the family, the family requested that Pat say a few words at Dale's memorial service.

Pat continues to think of Dale and the special relationship he and Ben shared. "One of the promises I had made, and unfortunately had not been able to keep before Mr. Clark's death, was for Ben to give cart rides to the grandkids. Ben kept that promise this year when we went to the family farm to visit. We spent a wonderful afternoon with Mrs. Clark [and her family]. We left with a deeper knowledge of how special the work was that Ben had done."

When Dale was buried, the family honored his request for three pictures to be placed in his chest pocket: photos of each of his two granddaughters and one of gentle Ben.

Chapter

Two

Tikva

Below are Cindy Ehler's journal entries of the time she spent in New York following the September 11 World Trade Center terrorist attacks, where she was involved in animal-assisted crisis response. This first-hand chronology provides several moving stories of how she and her Keeshond, Tikva, were instrumental in providing comfort. Cindy is the founder and president of HOPE Animal-Assisted Crisis Response (AACR).

FRIDAY, SEPTEMBER 21

Tikva and I leave Portland airport on American Airlines and arrive at La Guardia airport in New York. Kay, a mental health counselor and Delta Society Pet Partner, picks us up and drives us to her home, an hour north of New York City. We spend the night.

SATURDAY, SEPTEMBER 22

After much driving and walking through Manhattan we arrive at 55th and Westside Hwy., The Family Assistance Center at Pier 94. Red Cross volunteers tell us that we need to go to Brooklyn for processing. Many roads are blocked and the Brooklyn Bridge is closed to the public. We stop the van and ask information of a police officer. We say, "We were called out here from Oregon to help, and we need to go to the Red Cross in Brooklyn for processing. Could you please let us go over the bridge?" To my surprise, he allows us to go.

It takes us about six hours to process through at the Red Cross. Much of that time is spent letting the dogs visit with everyone. Kay, Josiah, and I are given our badges and hotel information. Afterward, Kay drives us back into Manhattan to our hotel. She has to return to her job at Yonkers Hospital for a few days. We promise to keep in touch and see her soon.

SUNDAY, SEPTEMBER 23

6:30 A.M. We arrive at the Family Assistance Center. Our names are not on the list of volunteers so we wait around about an hour before we are allowed to enter. Police officers and people from various organizations come over to see Tikva and Hoss [Hoss is Josiah's German Shepherd].

Families reach out to pat the dogs as we pass. There are other "therapy" dog teams visiting and we are informed by one of them that there is a visiting list. We look at the list and decide to go to Ground Zero.

I assume that since we are wearing our "HOPE Crisis Response AACR Team" badges, we are also Red Cross volunteers. Going on this assumption and thinking we already signed in when we got our stickers, we check out our decision to go to Ground Zero with a Red Cross chaplain who also thinks this is a good idea. She recommends that we get a police escort but doesn't tell us where to go. We don't think to ask either.

Cindy Ehlers (*in back*), Tikva, and Ground Zero emergency worker. (Photo by Josiah Whitaker)

Down at Ground Zero we make our way through heavy machinery, trucks, police cars, and other emergency vehicles, all coming and going in every direction through the powdery cement and ash-filled streets. Tikva and Hoss seem to sense the seriousness of the situation and follow our directions like well-trained soldiers.

We receive a wonderful welcome when we arrive at the Red Cross station located at the Independence School on Chambers and Greenwich Streets. Red Cross volunteers, police officers, and firefighters come over to see us and ask questions. Everyone thinks Tikva and Hoss are search-and-rescue dogs. Tikva senses the job she has to do and enthusiastically greets those who approach her. We're introduced to Karen Soyka, a Red Cross Disaster Mental Health (DMH) volunteer who we are to assist. She has never worked with dog teams. Some of the men are sitting in chairs, others are watching their comrades sift carefully through debris, hoping for any sign of life.

As we approach one of the many stations around "the pile,"

Karen is amazed at the difference in the firefighters with the dogs around. A look of gratitude sweeps over the dirt-stained faces of several men when they find out that these dogs are brought in to give them comfort and emotional support. Some joke, others ask what it is that the dogs do. I make my way toward the men sitting down and ask Tikva to go say hi. Her enthusiasm changes to calm and she is keen on making her way to one man sitting on a chair. She gently touches her nose to his knee, sits down and raises her paw up. Just like I taught her. As the man reaches out to pat Tikva, she gently places her front paws on his leg and stretches her head up toward his face. I beam inside, glad for the long hours I spent training her.

After I got back home and began the Red Cross training, I realized that maybe Red Cross works differently with outside organizations. (A couple of days later I found out that we were supposed to sign in at another booth.)

We remain at this site for about two hours. Conversation is

Emergency workers, Cindy Ehlers (center), and Tikva at Ground Zero. (Photo by Josiah Whitaker)

lighthearted with one firefighter, while another strokes Tikva's fur, sharing his experiences of the last twelve days. Sadness and grief replace shock and anger as each man shares his account of that first horrific day and the hours that followed. "Wow," says Karen in awe as we leave the site. "Those dogs did in a few minutes what it has taken me days to do."

Police officers pause from their tedious task of directing people away and asking to see badges so they can visit with Tikva and Hoss. As I watch our dogs brighten the gray gloom of sorrow and sadness of each person they come in contact with, I know this is where we belong.

MONDAY, SEPTEMBER 24

Josiah and I stop at the VMAT (Veterinary Medical Assistance Team) station before beginning our day at Ground Zero. We sign in at our station and meet up with Karen again. It takes about four hours to make it around the perimeter of the WTC site because almost every person wants to visit with the dogs. Both Tikva and Hoss appear calm and unshaken at the sights and sounds they are confronted with. They are gently intent on each person that they draw near to or are approached by.

"This dog sure makes my day," a police officer exclaims. People seem to viscerally feel the assistance, comfort, and emotional support that the dogs give. We know it when such words are echoed again and again by those we are there to help.

Passing by Liberty Street I glance quickly over to see the skeletal remains of one of the towers. The metal shards are rising over what used to be the Marriott Hotel. Crumpled metal melts over the sides of what once were walls over forty stories high. Smoke billows from the pile and I can faintly feel the heat rising from under my feet. I look away, feeling as if I had entered a forbidden, sacred burial ground. The scene is forever etched in my mind.

In the days that follow, I notice that every dog in the team looks sad and mournful when they pass this area. At one site in partic-

Josiah Whitaker, Hoss, Tikva, and Cindy Ehlers.
(Photo courtesy of Josiah Whitaker)

ular the dogs are dismal. They appear despondent and somber, wanting to be beside or near the workers but not interacting with them. I cannot explain why this is but I see the same expression repeated on every dog.

Months later I learn that the area directly under where the dogs were most despondent is where several bodies were uncovered.

TUESDAY, SEPTEMBER 25

Pier 94, Family Assistance Center. I think this is the first day the ferryboats are taking the families to the site of the WTC. We are asked by the mayor's office to be on the ferries and at the site with the families. I allow Tikva to lead the way among the families, not quite sure of what is expected of us at this point.

Once at the site, families began holding each other. Their open display of sorrow and grief makes it extremely difficult to control my emotions. I reach down inside myself, somewhere far away in my past, and gather the strength to keep the tears from spilling out.

The only way I can express my empathy and desire to comfort is through Tikva. I direct her to be close to the people who I think would want and benefit from her presence. She is a comfort to many on the ferry ride back to Pier 94. She walks down the aisle and, upon invitation, gets closer and then gently puts her paws up on their lap and snuggles in when they lean forward to hug her. A few minutes are spent at each table.

Many families comment that the dogs are a welcome sight. A woman who is sitting a couple of chairs in from the aisle reaches toward Tikva. "I can't believe you are here," she whispers. As Tikva draws closer, the woman hugs her, burying her face deeper into the fur on her neck, and begins to cry. "What's her name?" she asks. "Tikva" I reply. Looking at her friend and back at me, she mentions that Tikva means hope in Hebrew and she is Jewish. She makes the comment that her husband loved dogs more than anything except for her and that he had a dog—she thinks Tikva looks like his dog—before they were together. His bedroom cupboard is

Cindy Ehlers, Tikva, and Ground Zero workers. (Photo by Josiah Whitaker)

lined with pictures of his dog and he spoke often about him. She states softly that Tikva is a gift from her husband, a sign of hope to help her during this time of loss.

OUR LAST DAY AT GROUND ZERO

More and more workers are sent in from FEMA and they quickly make use of the "comfort" dogs, calling them over and asking them to get in their laps or jump in the golf-cart-type vehicles with them. I could tell by the downcast faraway look of one man that he was low in spirits. I sensed that he would welcome Tikva so I picked her up, carried her over and put her in his lap. He glanced up, his faraway look suddenly changing to relief. He sat there quietly, holding and patting Tikva. Tikva waited patiently. When I lifted her back into my arms he thanked me and asked me to bring her back again.

POSTSCRIPT

There was e-mail after we got home from a firefighter asking, "Where are the comfort dogs? Those dogs were the only thing that helped me get through the day."

Chapter Three

Bear

To protect the confidentiality of the child in this story no name is used. The story is a true account of events as told to the author by Shari Ferguson.

The dog had two strikes against him when Shari first found him. First, he was a Chow Chow, a breed known for its aggressiveness and difficulties with training. More importantly, he had been severely abused. Every rescue group that Shari contacted refused to take him. A Chow that had been through this type of abuse, they all said, would never be "adoptable" and would have to be put to sleep. But Shari's heart told her that this dog had something in him that was worth saving. The neighborhood kids decided he was like a little black bear cub, so Shari kept the name "Bear" and adopted him herself.

For many months Shari and Bear worked diligently together. Bear's neck had been so badly damaged by a wire clothesline that Shari decided obedience classes or pulling on his neck in any way was out of the question. Nonetheless, he learned "sit," "stay," and "down"—and how to behave with both dogs and people. Shari's two other dogs were already certified therapy dogs in their own right, well behaved with both people and animals, and they taught Bear what was acceptable dog behavior. To help Bear get used to being around people, Shari began taking him with her whenever she could. When she coached a local girls' softball team, Bear would sit in the dugout and kids would pet him. He not only got used to the attention but grew to like it. Bear, Shari found out, would gurgle when he was happy.

Probably because of his past abuse, boys and men were especially scary to Bear. With Shari's concentrated attention—pairing safe men and boys with yummy treats—Bear became less afraid of them, too. He was skittish at first, but he quickly warmed up to someone petting him or giving him treats. After he passed his temperament test Shari decided to let Bear try his skills at the Devereaux Children's Treatment Center. Many of the children at Devereaux had been through traumatic situations and Shari hoped that Bear's story would be the opening that some of them needed to start their emotional healing.

When Shari and Bear entered the children's treatment center for the first time, all the other dogs that had come to visit went easily to chase balls, go for walks, or just run around. Only Bear went straight for the corner and sat. He couldn't have missed seeing the small boy already sitting there, all alone.

At first, both Bear and the boy showed an uncertainty about what the other might do. A sideways glance here, a seconds-long stare there, was the extent to which either acknowledged that they were sharing the same space. Finally, the boy appeared to realize that Bear was not going to harm him, scare him, or even approach him if he did not want him to. So, hesitantly, he reached out his hand and lightly touched Bear's abundant fur. His fur was so soft it felt like touching air. It made a delicate tickling sensation on his palm and his fingers gradually worked deeper into the warm fur. Then slowly, the little boy began repeatedly stroking Bear's velvety head, which progressed to rubbing Bear's ears. The boy found that the big dog continued to sit passively, making a slight, funny noise in the

Mary (*left*), Bear, Denise, and Tamu. (Photo by J. Pamela Culpepper)

back of his throat. He even seemed to smile. The little boy wrapped both his arms around the big dog's neck and buried his face there.

For Bear's part, the boy's tentativeness was calming. Bear had been with one other boy before and that boy had shot salt pellets at him while Bear was tied with wire clothesline to a tree. *This* boy, Bear seemed to decide, was not like that awful boy. This boy was nice. Over the next hour, Bear's boy continued to hug him. He hung on him mutely, comforted by Bear's quiet presence. Bear also felt relaxed by the gentle presence of the boy. He returned the favor by licking the boy's ears and face. The boy didn't mind and even seemed to enjoy this attention. Both were having a good time.

All of this went on relatively ignored by everyone else. There was nothing to draw attention to Bear or the boy. The other children were mostly uninterested in a dog that didn't do anything. For Shari, though, there was a sense of relief. Bear had found a niche for himself. He was not the kind of dog that liked to fetch and play with Frisbees. But, as this visit confirmed, Bear was the kind of dog that was quite content to just sit and be stroked.

As she watched the two in the corner, Shari reflected on the events that had led Bear to her that fall in 1994. She and her dad, Dick, had been on a walk with her two dogs, Tamu and Reveille, when they came upon a little black Chow. Shari could see that he had a wire clothesline wrapped around his neck and that he was bleeding from several different places on his body. The Chow watched from a distance as Shari tried everything she could think of to show him that he was safe. No matter what she tried, though, he would not come close enough to be caught. Shari and her father gave up and continued their walk, only to notice that the Chow was following them at a safe distance. However, by the time they had returned to Shari's home, the Chow had run off again.

Over the next week the Chow continued to elude them. Living next to a busy intersection as she did made Shari afraid for the Chow. She prayed, "Dear God, please help me find a way to help that Chow." Shari would never have guessed the turn of events that would answer her prayer.

If there had been no construction around her home, there would have been no need to tie Reveille to the six-foot ladder. If there had been no cat to chase, the ladder would not have fallen. If the ladder hadn't fallen, Reveille would not have been scared and run. If he hadn't run, the ladder wouldn't have chased him.

Reveille was trying to figure out how to get away before the ladder attacked him, but the faster he ran, the faster the ladder kept coming. There was such a racket from the ladder scraping the concrete that people from every corner joined Shari and her dad, trying to rescue Reveille. A hive of frantic people, a dog, and a ladder turned onto the busy four-lane road, heading toward oncoming traffic.

Shari would say it was at that moment that God's angels sprang into action. Somehow, all of the vehicles managed to miss Reveille, the evil ladder, and the chasers, who were by this time far enough behind Reveille that he couldn't hear them yelling at him. Eventually, they became so tired they couldn't yell, much less catch him. Reveille gave up on the ladder and somehow figured out a safe way home while Shari and Dick were still out searching for him. Shari's neighbors spotted him and Reveille was finally rescued from the wicked, clattering ladder.

Shari and Dick walked around the back of the house, and a small noise made them turn. Standing on the patio was the little black Chow. He had heard all the commotion caused by Reveille and the ladder and had followed him home. Shari's dad quickly

Shair Ferguson surrouned by Bear, Tamu, and Reveille.
(Photo by J. Pamela Culpepper)

closed the door between the patio and the garage, the only way of escape. Now the task of showing Bear he was safe would be easier.

Bear's trials did not end when Shari and her dad were finally able to remove the wire from around his neck. Shari's vet found, in addition to the abuse caused by the pellets and the wire, that Bear had heartworm, other internal parasites, and problems with the tear ducts in both of his eyes. He estimated Bear's age at around two years but cautioned her against getting attached to the Chow. "He probably won't make it." It was too late; Shari was already attached.

The boy who sat in the corner with Bear, Shari later learned, also had been physically abused. The withdrawal that Shari observed was how he always behaved and no one at Devereaux had been able to draw him out. He had resisted all the treatments, all the other children, and all of the therapists since his arrival. His connection to Bear was the first time that he had allowed anybody to get close to him. Every visit thereafter, Bear and this little boy would spend the entire hour together, their bond growing closer with each visit.

Shair Ferguson and Bear. (Photo by J. Pamela Culpepper)

People openly remarked how the more they were together, the more outgoing they both were. Not only was Bear helping to heal the boy, the boy was helping to heal Bear. In the boy's therapy sessions, his therapist was able to help the boy talk about how Bear might have felt by telling him about Bear's ordeal. In this way, the boy could express indirectly how he himself felt when he was abused.

Bear has been a therapy dog and a regular visitor at Devereux Treatment Center in League City, Texas, since 1995. Along with the other animals in "Faithful Friends," he goes to several units at Devereaux on a monthly basis. Many of the kids there know his name and call to him as soon as he gets out of Shari's truck. They also know Bear's story, how he came to Shari, abused and scared, and slowly learned to trust. They quietly pass the story on to the new kids who come to stay and warn each other to be careful of his feet (Bear is still skittish about anyone playing with his feet after all the needles and tubes that he had at the vet's office). And Bear is still quite content to just sit and be stroked by the children who need his quiet presence.

Chapter Four

Molly

That tenth morning Shannon opened her eyes to the same still life that was there when she had lapsed into sleep a half hour before. In the foreground were the blurred red and gold wisps of her own wavy hair laying against the scruffy recliner. In the middle ground stood the functional hospital end table burdened by a clunky institutional phone and standard issue lamp. Bits of paper with scribbled phone numbers were scattered over and under it. Shannon's hazy vision could make out in the background her daughter's blonde jewels of hair. Instead of the long waves that had floated past her seat as she rode her new bike, there were now haphazardly chopped and spiked chunks of blonde hair splayed over the stiff whiteness of the hospital pillow. Shannon's heart felt lumpy and dead in her chest.

She's just the same, thought Shannon. The thought was not a comforting one.

Shannon's last ten days had been spent glued to Hannah's ICU bed, listening to the hum and beeps of the machines whose tubes and

Hannah Redford and Molly, 1995. (Photo by Lee Gaffney)

cords threaded into and around her eight-year-old daughter, and watching the steady stream of nurses, doctors, and respiratory therapists. She was unable to accept the awfulness of what had happened.

Hannah had been whipped thirty-seven feet into the air by a hit-and-run driver. She was not, thankfully, dragged and bumped several more blocks under the car like the heap of pink rods that used to be her bicycle. When the doctor left the operating room to talk to Shannon, he had not minced words. The cervical nerves leaving Hannah's brain stem had been severely dislocated, leaving her without the ability to regulate her own vital signs. Her chances of surviving through the night were small.

Shannon clung to any fact that helped her continue to hope. Her biggest relief was that her daughter had not died that first night in the intensive care unit. In fact, Hannah had survived nine nights in ICU, despite a body temperature that fluctuated drastically and oxygen levels that took periodic nosedives. She had spent last night on the pediatric floor. Shannon tried to think of this accumulation of facts as mounting evidence of her daughter's journey to wakefulness. But nagging little doubts kept pushing their way in and her mental and physical exhaustion was making it hard not to give in to the despair. She escaped to faith and fervently prayed for her daughter's life: Dear God, let it be thy will that I have not seen the light in her eyes and dimples of her smile for the last time.

The Visiting Pet Program was just then beginning their regular Saturday morning circulation through the pediatric floor. Lee was leading Molly, a little white terrier with wiry hair, into Hannah's room when a floor nurse stopped her. Quietly the nurse discouraged her from entering. "There isn't much point in visiting in this room. The little girl is in a coma."

Shannon jumped to her feet. Her fears and hopes were bumper cars banging and twirling in her head. This is a clean and sterile environment! she thought. Why would anyone bring animals in here? But

Hannah Redford and her mother, Shannon Cote, training their dogs to heel. (Photo by Sharon Ruthven)

Hannah Redford at her church library (Photo by Sarah Ruthven)

she recalled how much Hannah loved animals and waved Lee and Molly into the room. Shannon looked down at the little dog with the huge, round, brown eyes and the wildest white fuzzy coat she had ever seen, took a deep breathe, and instructed Lee where, on Hannah's bed, to place Molly. Lee lifted Molly onto the bed near Hannah's face, then moved Hannah's hand, brushing it against Molly's hair. "Here's Molly, Hannah," Lee said. "Can you pet her?"

Lee continued working with Molly and Hannah as other members of the Visiting Pet Program softly entered the room. They stood awkwardly watching Molly and Lee, asking Shannon about Hannah: What was wrong with her daughter? Shannon, tired and distracted by the questions, had turned to respond when she heard someone say, "Look, she's petting Molly!" Shannon whirled around and zeroed in on Hannah's small hand. It wasn't much, but it did appear that Hannah was moving her fingers.

Molly was just then very gently licking Hannah's face, as though she was trying to wake her. Lee alone, leaning close to Hannah and holding her hand, saw that Hannah's closed eyes had begun to flutter. It looked like a newly born butterfly attempting to open its wings. Lee watched intently as Hannah's eyelids continued to quiver, never quite opening, and decided to intensify the demands on the little girl. She moved Molly above Hannah's head and said, "Find Molly, Hannah. Can you pet Molly?" Shannon watched, afraid to look and afraid not to.

When Hannah's arm began jerking upward to her new friend Molly, Shannon gasped and the tears started flowing. Screaming with joy, she grabbed Hannah's nurse. The two elated women held tightly to each other, jumping up and down like pistons in an engine.

"At that moment," Shannon recalls, "I knew that my baby girl was still in there and that someday she would wake up."

Hannah awakened slowly over the next two months, a normal course for people who have been comatose. Molly's initial triumph opened the door for a whole zoo of animals to participate in her recovery, including a pot-bellied pig and a slender wiggly fellow that Hannah insisted was a weasel.

Molly with Lee Gaffney, 2002. (Photo by Mitchel Osborne)

Shannon recalls,"I tried to tell her that there was never a weasel and she argued with me, saying 'Uh-uh! They put a weasel under my gown and it tickled my tummy!' Hannah's older sister started laughing. She said, 'Excuse me, Mom, but I think Hannah is halfway right. They did put Weasel under her gown, only Weasel was really a ferret.' "

On June 2, 2003, Hannah will graduate from high school, ten years to the day after the hit-and-run driver launched her high into the air and dragged her bike to raggedy shards of metal. Although she has a noticeable jerk to her steps and her speech is somewhat labored, Hannah has every right to walk with head held high across her high school stage to receive her diploma. Hannah and Shannon have great hopes that their friends, Molly, Lee, and Tommy Gaffney, will be there to watch.

Lee's fuzzy little white mutt is now sixteen years old. Molly started her life with Lee and Tom as an abandoned dog from the New Orleans SPCA and recently passed her yearly physical with a recommendation for continued therapy work. The opinions of the staff at Oschner Hospital are that Molly is a consummate professional who always knows who needs her special attention.

Shannon knows that Molly not only helped Hannah but also helped her continue to hope at a time of little hope. "As a mother," says Shanon, "I will never be able to thank Lee and Molly enough. [Molly] is an angel with four legs and a tail."

Chapter Five

Marley

By the time she graduated Melissa Gordon knew for certain that she wanted to partner with an animal in her work. She had learned about animal-assisted therapy in her graduate Occupational Therapy Program at Shenandoah University and knew how people working with animals made greater gains than people who received traditional occupational therapy. There really was no question in Melissa's mind about the benefits of animal-assisted therapy. The question was rather how to show others those benefits. When she was hired by the Program for Students with Physical Disabilities in Montgomery County Public Schools, she began putting together the pieces of her future.

All kinds of animals—dogs, cats, rabbits, even llamas—have been certified as therapy animals. For Melissa and her husband, Joe, the joy that Golden Retrievers exuded convinced them that this was the kind of animal for them. As luck would have it, Joe's colleague from work was breeding his female Golden to a champion male and agreed to reserve a puppy for the young couple.

The puppies—four males and four females—were born on April 3, 1999. Eight weeks later, on the day they went to choose their fuzzy ball of fur, Melissa fairly bounded out the door. She had been reading about testing the puppies' personalities using the Puppy Aptitude Test and knew what to look for in a therapy dog—a puppy that would follow, one that was friendly and not overly sensitive to touch or sound, and was smart. The dog she chose was Marley.

Marley was a happy-go-lucky learner that graduated from puppy training to basic obedience and then to advanced obedience and agility classes. Agility courses are like playgrounds for dogs. They have tunnels and ladders and jumps and tires and poles that dogs go through and over and around with cues from their owners. Marley loved this! Just after his first birthday, Marley received his Canine Good Citizenship certificate. This confirmed that Marley was a well-behaved little fellow that could adapt to a variety of people, other dogs, and unfamiliar situations. Melissa then enrolled Marley in a therapy dog course. Again, he did not disappoint; he earned his therapy dog certification just a few days before his second birthday. Two months later, Marley began work as Melissa's occupational therapy partner at Laytonsville Elementary.

The first morning that Melissa walked Marley into the classroom the children reacted with excitement, smiling, laughing, and chattering all at once. Five-year-old Jaison, in particular, was very excited to see Marley. Jaison was scheduled to be Melissa's first O.T. session that day. Jaison couldn't talk; his responses were either laughing or crying. He was usually not willing to work with Melissa. His spastic quadriplegia made it difficult for him to hold objects and when she worked with him to improve the functional use of his hands, he frequently cried. But Jaison's eagerness to be around Marley flooded her with hope. A few minutes more of group greeting with the classroom kids went by before Melissa began Jaison's fine motor skills routine. It was quickly apparent, as she had hoped, that Jaison was not going to resist. "In fact," said Melissa, "he was so excited to work with Marley, he actually enjoyed stretching out his arm to pet and brush the dog."

Jaison Johny, Marley, and Melissa Gordon during Jaison's occupational therapy session. (Photo by Dick Dressel)

Buoyed by Jaison's initial upbeat response, Melissa continued her demands. Jaison next worked on brushing Marley's teeth (a favorite of Marley's) with Melissa helping him hold the brush and move his wrist. Jaison barely noticed that he was now vigorously using muscles that he rarely moved. Melissa and Jaison's next task was to throw the ball, an exercise that broadened Jaison's range of motion and worked on the grasping and releasing of an object. After a throw, Marley did his part by "throwing" the ball back to Jaison. Jaison chuckled and giggled the whole time. The trio continued to work on new activities and Jaison was eager to keep going. Sprinkled in between the exercises were Marley's rewards for the "work" he was doing. Jaison held Marley's treat and Marley gently nudged Jaison's hand open to retrieve it. The action of his nose against Jaison's palm helped relax Jaison's grasp reflex. The therapy session passed quickly as Marley and Jaison connected like electricity. The usual sixty minutes ticked on to seventy, eighty, ninety minutes, with Jaison showing no awareness that he was working and working very hard. He cried when it was time to go.

Over the months that Jaison worked with Marley, Melissa saw enormous gains in his muscle tone, coordination, and endurance. His occupational therapy session eight months later showed that Jaison could maintain a grasp—something he had not accomplished before working with Marley—could use his right hand to hold the toothbrush and the hairbrush, and could make larger movements on his own to brush Marley. Jaison could reach for an object with either his right or left hand and transfer it to his other hand on his own. By himself he could throw a ball (both grasp it and release it). Because Melissa and Marley had also participated in Jaison's physical therapy he could now hold Marley's leash while he walked in his gait trainer. Melissa was ecstatic! Perhaps the most exciting thing that happened wasn't even in her area of expertise. Jaison talked.

"Jaison had begun uttering short words earlier in the year ('hello,' 'bye,' 'mommy,' 'yes,' 'no') but not too many words in context. As we worked, Jaison called for Marley, stating his name perfectly. I worked with Jaison to give Marley commands such as 'sit,' 'down,' 'come,' and 'catch ball.'" Said Melissa, "I was amazed! Jason even called Marley, said 'cookie' as he gave Marley a treat, and told Marley he was a 'good boy.' Jaison had never said any of these words before."

Since Marley and Melissa became a team, they have seen many kids achieve similar levels of success and beyond. Kids are more willing to work on difficult tasks and get much better when a certified therapy animal is included in their therapies. Without the therapy dog (it is often a dog) a child is afraid of the muscle aches that range-of-motion exercises bring. With the therapy dog present there is purpose to the actions; aches are hardly noticed. Without the therapy dog a child may be happy with sedentary play. With the therapy dog kids challenge themselves to do difficult movements (like walk with a gait trainer). Without the therapy dog a child may not care about learning new words. But with a trained and certified dog there is something wonderful, exciting, and

Jaison Johny, Marley, and Melissa Gordon. Marley provides the incentive for Jaison to stretch contracted muscles. (Photo by Dick Dressel)

special to tell and the need for the words to tell it. The temptation is to assume that Jaison's results are unique and not reproducible. But Melissa can name many other children who have shown similar leaps in skills.

Christian was one of these children. A nine-year-old student with mental retardation and visual and speech impairments, Christian did not like people in his space. Nor did he like to be touched. At home he often shoved aside his baby brother when they sat close together, a fact that troubled Christian's mother. That first day that Melissa brought Marley to school, Christian's mother, Rose, was there to watch. She was excited about animal-assisted therapy but was concerned over how Christian would react with an animal near him.

Christian Braisted–unsure of allowing Marley next to him. Rose Braisted in the background. (Photo by Melissa Gordon)

Melissa began her session with Christian by having Marley lay down on the floor and encouraging Christian to sit next to him. Christian would have none of it. He began to cry and pushed Marley away. Marley remained laying, unperturbed, where Melissa had commanded him to be, on his side, with his eyes closed. Melissa watched the two closely. Like Rose, she was concerned about what Christian might do, but also curious. Over the next five minutes Christian sat next to Marley, whimpering and occasionally shoving him. Then he stopped. Still looking doubtful about allowing the big dog next to him, Melissa could see that Christian was watching Marley's golden coat move up and down, watching his abdomen rhythmically expanding and contracting as he breathed. Christian crept closer to Marley.

In the space and time between this movement and the next, the world slowed. Marley remained still. Melissa and Rose remained silent, trying not to breathe. Christian watched the up-and-down movement of Marley's trunk. . . .

When Christian next moved, it was to place his forearm along

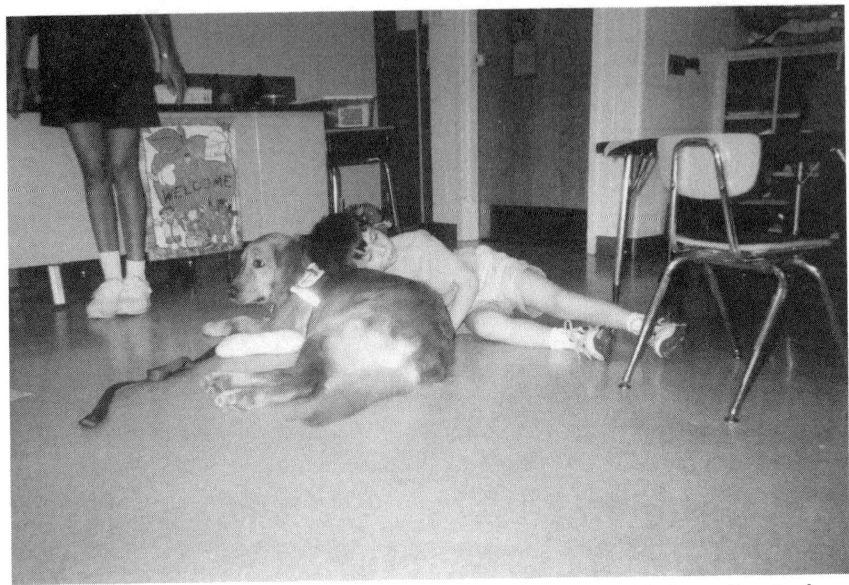

Christian Braisted finally relaxing with Marley. (Photo by Melissa Gordon)

Marley's back. Now he could feel Marley's breathing as well as see the movement. Marley still lay resting peacefully on the floor. The dark-haired boy's eyes closed and his frown relaxed. He lowered his head and body onto Marley's back. One arm relaxed across his own chest; the other was still draped across Marley, monitoring the rhythmic movement and heartbeat of the warm dog. Marley was unperturbed.

As Christian lay, listening to Marley's heartbeat, his mind balancing the lub-dub lub-dub of Marley's heart with his own heart's pulsing, he next began softly patting Marley, matching the beat of Marley's heart to his pats. Rose and Melissa stared at each other, knowing something very special was happening. The rest of the classroom staff were equally amazed by the events unfolding in front of them. Everybody knew that Christian did not touch people. How striking it was to see him so obviously connecting with Marley. For twenty more minutes Christian and Marley continued to lay together, Christian stroking and patting Marley, Marley laying stretched out and placid. At the end of the time, when his vision therapist came to escort him to her room, Christian rose without complaint.

After Christian left, Melissa began therapy with another child. She had been working for about fifteen minutes when she heard someone running down the hallway toward the classroom. Flustered and excited, the vision therapist rushed through the doorway toward Melissa. Christian, she announced breathlessly, had just said his first word.

Marley's training allows children to interact with him in safe ways that motivate them to try new, fun things that advance their skills. His gentle temperament, his eagerness to please people, and his acceptance of people—some say his ability to be present in the moment—produce results that do not happen without him. If Melissa alone, or any of us for that matter, achieved the results that Marley has, people would be knocking down our doors with pleas for us to help them.

Perhaps one day it will be common practice to include animal assistance in occupational, physical, emotional, and other therapies. And perhaps one day the results that Marley and Melissa together achieve also will be commonplace. Nothing else achieves these results so quickly. But Marley, with the help of his partner Melissa, does so regularly, and all for the price of a handful of treats.

Chapter

Six

Brutus

Five years ago Loretta Brobst was desperate. Intertwined with her morbid obesity were substantial health problems and chronic, severe psychological problems. Depression was her constant companion. She often did not want to get out of bed and, when hospitalized with a medical complication to surgery, did not care if she lived. The functional problems that came with extreme obesity were daily frustrations—tying shoes was impossible and walking for long unthinkable. She had lost her job as a registered nurse. Every day, people on the street, as well as those whom she thought she could trust, had roughly sliced her with their words. The derogatory comments had burned deep into the center of her fragile spirit. She was vulnerable and did not want to hear those wounding words again. She started isolating herself, thinking that people were repulsed by her more than 400 pounds.

The emotional harm that others caused her was the worst part of being obese. And Loretta is no different from anyone else with extreme obesity. Research from the National Institutes of Health

Brutus shares Loretta Brobst's hospital lunch. (Photo by Benedict Gress)

shows that the psychological burden is the greatest adverse effect that morbidly obese people must deal with.[1] Clearly, Loretta would be better off in every way if she lost weight. So why, if she was so aware of the problem, did she not only *not* lose weight, but gain it?

Loretta's psychiatrist, Scott Berman, knew that Loretta's depression was deeply intertwined with her obesity and that the antidepressant medication she was taking was not helping her enough; she also needed therapy. Unfortunately, her anxiety over meeting new people had prevented her from starting this process. From her doctor's perspective, the secret lay in Loretta working with an animal-assisted therapist. He knew from firsthand experience the calming effects that animals have when one is highly stressed. Scott had had a rare neurological disease requiring a central venous catheter, surgically implanted. It had been a time of high anxiety for Scott. When Brutus came to the hospital Scott was able to relax, to sit and play with him. Scott knew that other patients from his practice had benefited from animal-assisted therapy and he had read the medical literature supporting its effec-

Brutus, Kayla, and Kathy Gress visit Loretta in the hospital. (Photo by Benedict Gress)

tiveness. He also knew that Loretta had grown up around animals and that she felt she could trust them. If there was an animal around to help her overcome her fears, he reasoned, Loretta stood a better chance of staying in therapy. "Go see Kathy Gress," he counseled her. "She has animals in her practice. Bonding with an animal would be an excellent avenue for you."

Kathy, a certified nurse psychotherapist, had a lifelong love of people and animals. In her private practice she had noticed that people of all ages, but kids especially, opened up to her when her pets were around, telling her things they had told no one before. When she looked for the research to support the use of animals and found mostly uncontrolled studies, she decided to do a study herself. Her master's thesis showed that college students had decreased biological markers of anxiety while in the presence of certified therapy animals.[2] The logical next step for Kathy was to train and certify animals whose natural inclinations answered the needs of people in therapy. Kathy was drawn to Great Pyrenees, a giant breed of dog with long white hair and a gentle nature. She

Kathy Gress's therapy dogs, Brutus and Kayla. (Photo by Benedict Gress)

read all she could about them, found a set of breeding dogs whose temperament seemed ideal—both were certified in animal-assisted therapy—and, in 1994, drove to North Carolina in the middle of a snow storm to pick up her new puppy.

Brutus was smart, affectionate, playful, eager to please, devoted and, most importantly, sensitive to people's needs. At six months, the youngest dog at the certification trials, he received his Canine Good Citizen certificate, an American Kennel Club designation. At ten months he earned his initial therapy animal certification through TDI (Therapy Dogs, International). At one year he earned his second national certification through Delta Society, Inc. These were remarkable acheivements for such a young dog, but they only verified what Kathy already knew: Brutus could give people the love and support they needed.

When Loretta conquered her fear of meeting people and had her first therapy session, she saw a therapist and a dog that

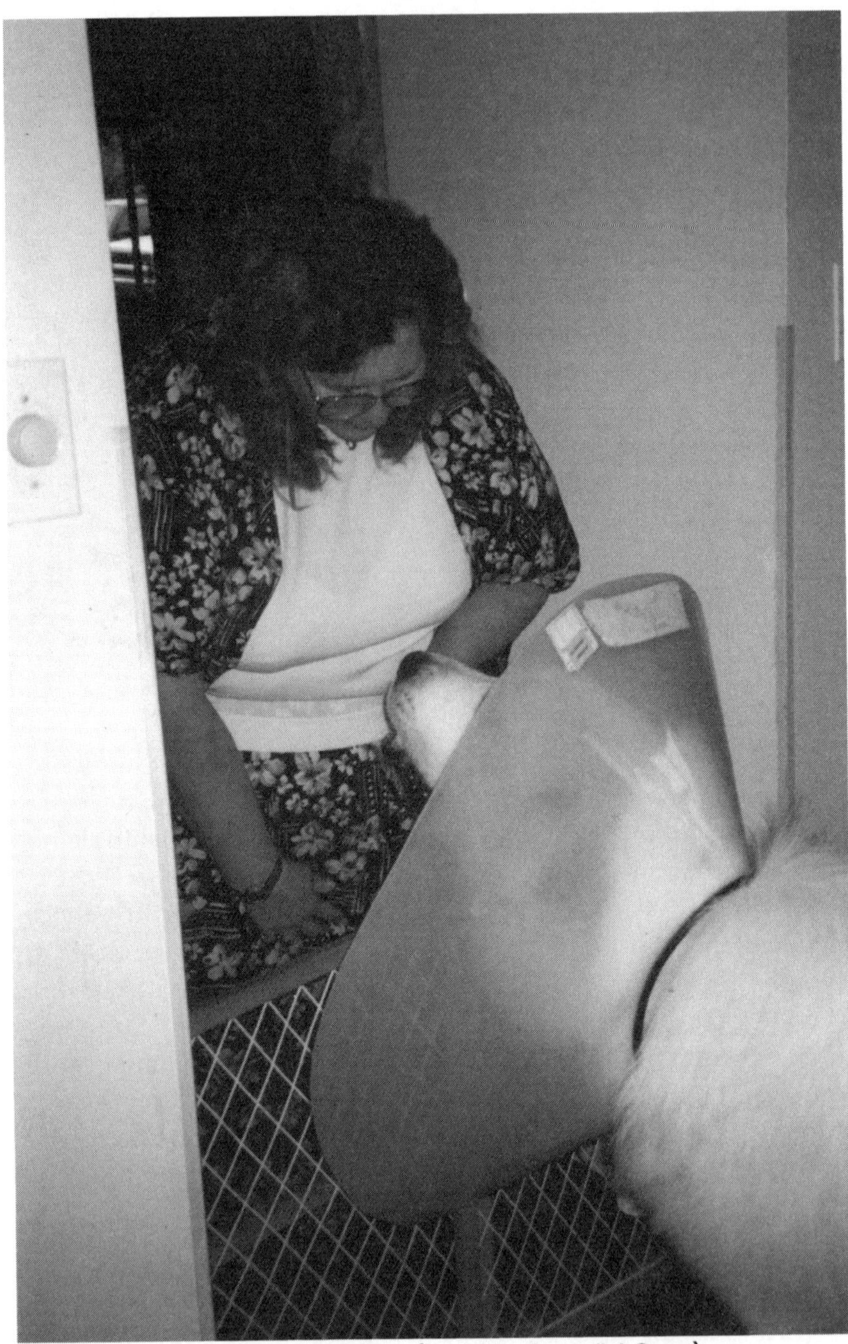

Loretta visits Brutus after his surgery. (Photo by Benedict Gress)

believed in her. "The very first time I met Kathryn and Brutus, I felt a closeness to them," remembers Loretta. "Brutus [accepted] me as a whole person and warmly [greeted] me each meeting."

In seeing that, week after week, Brutus accepted her, all of her, Loretta was learning to accept herself. Once she could accept herself as she was, she was ready to tackle more difficult therapeutic tasks. Loretta began to work on issues related to her depression. "I was not dealing well with depression, brought on by a bad marriage and weight problems. I would see Kathryn for weekly therapy sessions with Brutus. Brutus was always there for me."

Brutus always sat next to Loretta during her therapy sessions, providing the emotional safety that Loretta needed to confront difficult realities. She often stroked him as she talked to Kathy about the enormous psychological burdens of her obesity. The very act of stroking his soft fur helped calm her. "Animals," she said, "care about me no matter what. Whether you are fat or skinny, there is nothing like an animal to help lift you up. They make you feel loved when no [person] does."

Loretta's obesity continued to be her primary issue in therapy, one that caused her grave health problems. She was hospitalized with a severe leg infection. "Brutus and Kathy came to visit me. Just seeing Brutus standing at the door with his country western hat on made me feel warm all over. Brutus looked like he really did care about me. He was truly concerned about what was happening to me. I cried with happiness because Brutus and Kathy took the time to visit. That really meant a lot to me."

Over the course of the last two years Loretta has had multiple hospitalizations and many health complications. Each time Brutus was her motivation to keep trying. Following one of her surgeries, her doctors felt Loretta's health was worsening, that she was near dying. Loretta remembers that the doctors were counting on Brutus to give Loretta the desire to get well: "'Brutus is coming down the hall and he will help Loretta turn things around,' they said."

After five years of therapy Loretta has lost 215 pounds. She now weighs 196 pounds. While that figure does not suggest someone who is svelte, compared with the 400-plus pounds she started with, it is a record of the incredible work she has done. As a result of her therapy with Kathy and Brutus, Loretta has successfully overcome a life-threatening weight problem, has learned to be assertive, can better handle conflicts, has better interpersonal relationships, and feels happy with her overall better quality of life. There is one more surgery left, to eliminate the excess skin from her weight loss. Once she is done with the surgery, Loretta wants to begin working with Kathy and Brutus on her career goals. She is hopeful for her future. "I know that . . . Brutus and . . . Kathryn Gress will be there every step of the way. . . . Brutus is my driving force, a gentle giant with a guardian angel heart."

NOTES

1. National Institutes of Health, "Health Implications of Obesity." Consensus Development Conference Statement, February 1985.

2. Kathryn Gress, "Pets as Mediators of Therapy" (master's thesis, Kutztown University, Pennsylvania, 1996).

Chapter
Seven

Rio, Mandy, & Kitty

Laurie Hardman has three Portuguese Water Dogs: Rio (he "sings"), Mandy, and Kitty; each is a certified therapy pet. Together they have been visiting patients at Swedish Hospital in Seattle for nine years. As we accompanied them on their rounds, they greeted Doris Dolan, a patient with a thick German accent whom they had met the previous week and who had suffered a stroke. She found all the dogs a great treat. "These dogs are so smart," Doris marveled. "I think they should go to *all* the hospitals!"

Their next visit was with Bert Bruce, who was surprised to find Laurie's wonder dogs calmed the stress and pain he felt after a serious heart surgery. "She reminds me of my dog at home," said Bert quietly, while gently stroking the tranquil Mandy as she lay across his lap.

"Mandy always waits for permission before joining someone on their bed," says Laurie, who recalled Mandy's gentleness with another fragile patient from the previous week.

Laurie had been asked to visit a young man named Jim, who

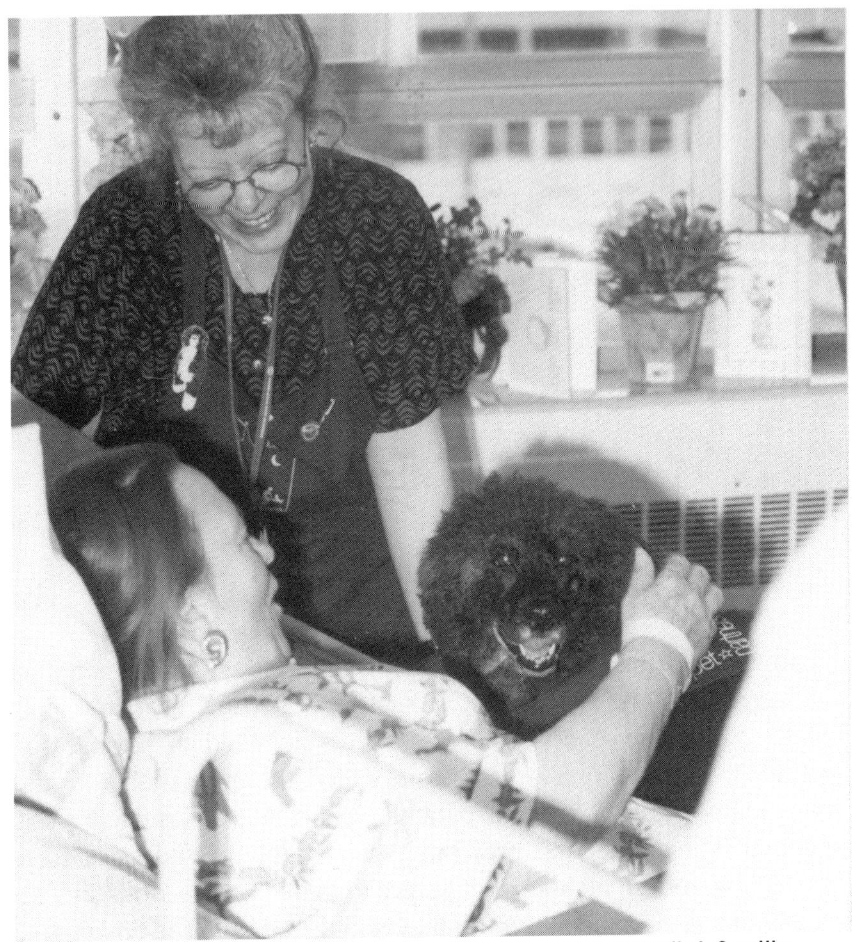

Doris Dolan, Laurie Hardman, and Mandy at Swedish Hospital, Seattle. (Photo by Donald W. Smith)

lay in the hospital with a halo (a rigid head brace) and a full body brace, waiting for a visit from Mandy. He had been unable to move anything besides his fingers and his face since his accident.

"Jim was extremely eager to have Mandy on the bed although, I have to admit, I was somewhat intimidated by all his equipment," Laurie recalled. "I took hold of Jim's right hand, and together we inventoried his bells and whistles until I felt comfortable. When the nurse indicated that Jim was ready, I gently placed

Mandy, Kitty, and Rio, certified therapy pets. (Photo by Donald W. Smith)

Mandy across Jim's body brace." Mandy immediately began nuzzling Jim's neck and Jim reveled in the attention and tender touches. He lay still in his cast, allowing Mandy to snuggle next to the one place on his body where there was warm, open skin. Then Jim struggled to move his arms up over Mandy's back. It took only seconds, but the seconds were heavy with Jim's concentration and exertion. Still, no one could turn away or take away from Jim the chance to do for himself this one small act. There was a collective deep exhale when he succeeded.

For the next twenty minutes, Jim gently stroked Mandy's curly side. His head brace kept him from seeing the tears that were gently rolling down his father's cheeks.

As the minutes ticked by to the time when Laurie needed to go home, Jim asked her if she would hold Mandy up for him to see. Laurie moved quickly to hold Mandy in front of Jim's face. "As I held Mandy up in front of him, he looked me directly in the eyes and simply said, 'Thank you.' 'No Jim,' I replied, 'Thank *you*.' "

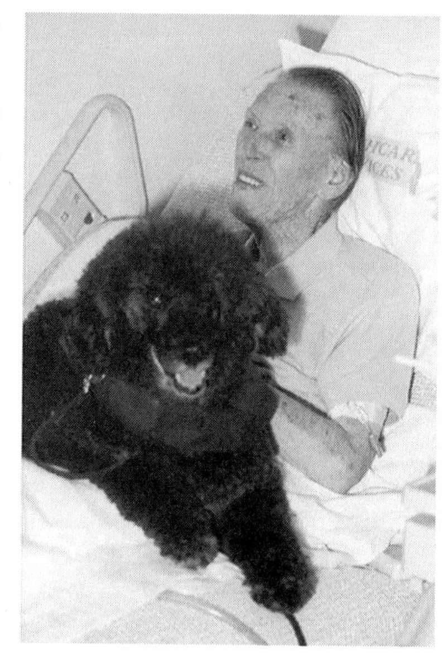

Mandy visiting Bert Bruce. (Photo by Donald W. Smith)

The next week when Laurie came to the rehabilitation unit, Jim had gone home, but Laurie remembers Jim's gift to Mandy with a smile. "The best gift, she said, "was when Jim christened Mandy's way of draping herself across someone's chest as "the melt."

"About six months later," Laurie continued, "my husband and I were waiting at the [hospital] elevator with Mandy and Rio, when a nurse came up to us. She said, 'You don't know me, but I know you,' and proceeded to tell us that she was Jim's mother."

Laurie learned that Mandy's visit had done more for Jim than start this seriously injured person believing that he might still have a future. At the time, he had not only just broken his neck, but also lost his own very special dog. Mandy's "melting" across his chest gave him the strength to face all the grief, fear, and pain that he was feeling, and gave him the resolution to move forward with his therapies and to heal both physically and emotionally.

"All this was wonderful news," Laurie concluded, "but the sweetest music was hearing Jim's mother smilingly report that Jim had completely recovered from his injuries."

Chapter

Eight

Taffy

It was fate when Taffy, a painted pony in her late twenties, was rescued from a barn fire. At the time of the fire Jennifer, her owner, was readying herself for college. It took two years for her to decide, with her mother's help, that Taffy would be happy being a therapy horse. Jennifer and her mother called Meggan Hill, horse trainer and North American Riding for the Handicapped Association (NARHA) certified Advanced Therapeutic Horseback Riding Instructor. Taffy passed all the exercise drills, veterinary examinations, environment and behavioral tests and horse profiles with flying colors, something, according to Meggan, only about 1 percent of horses reviewed can do.

Meggan remembers the day Taffy came to Cowboy Dreams. "She had a look, a sparkle in her eye. She was definitely special. Her eyes seemed so kind, as if she was asking, 'How can I help?'"

In hippotherapy a child and horse work together with the assistance of a knowledgeable therapist. Riding a horse moves the rider's pelvis, legs, and trunk in a rhythmic and repetitive way. The

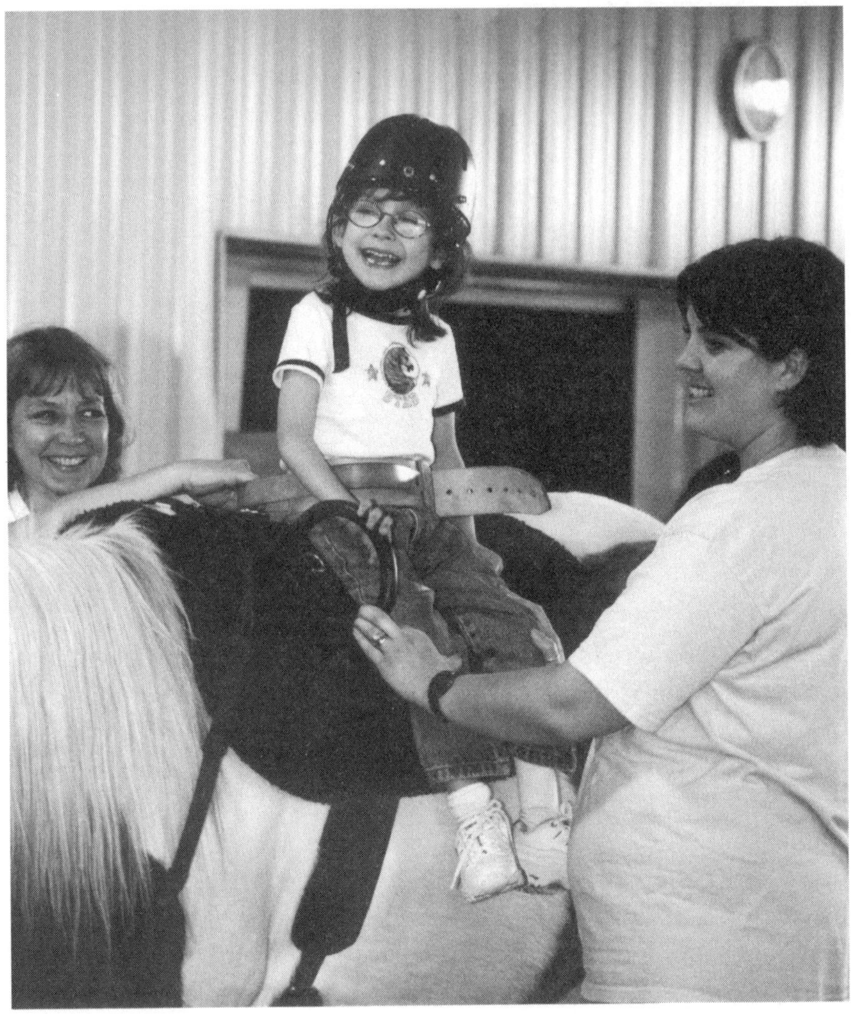

Caitlin on Taffy, with sidewalkers. (Photo by Cathy Lockhart)

horse's walk provides the rider with essential sensory input that simulates the human gait. With children who suffer from muscular disorders, the horse's body warmth reduces muscle spasms and increase's the child's hip and leg flexibility. The child's nervous system assimilates the information this movement provides, resulting in many significant, sometimes amazing, sensory and motor gains. A regular program of hippotherapy gives children

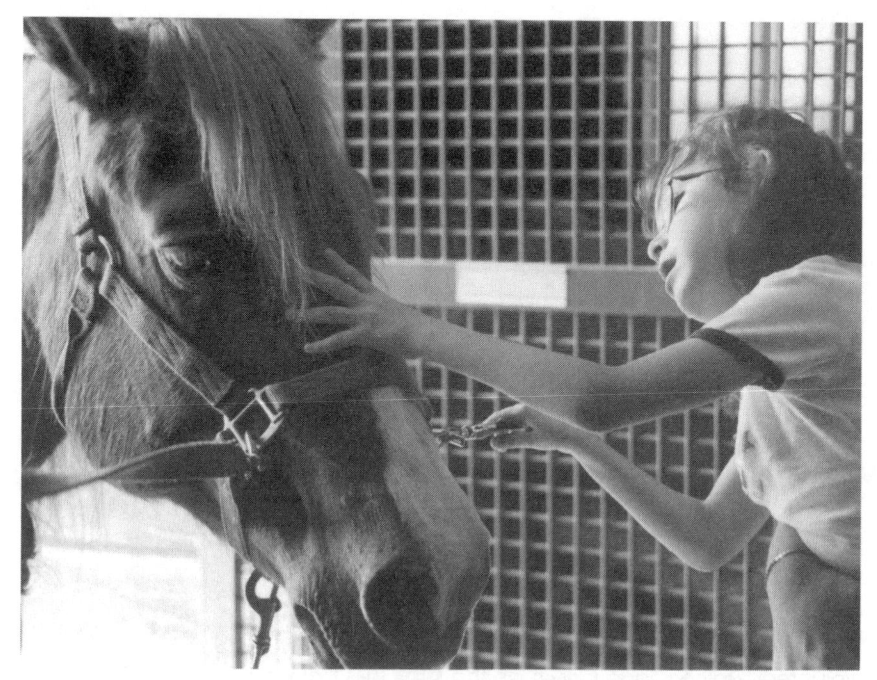

Caitlin readying Taffy to ride. (Photo by Cathy Lockhart)

notable improvements in mobility, strength, function, and coordination. There is no machine, no human, and no team that can offer the same benefits.

This was what Cari Oliver read in the local paper about Cowboy Dreams. Her daughter, five-year-old Caitlin, was born with quadriplegic spastic cerebral palsy and was asthmatic. She had already had brain surgery to repair a malformation. Her immobility left her dependent on her family to dress her, brush her hair and teeth, and feed her. Numerous medications and surgeries partly decreased her muscle spasms. The casts on her arms were intended to one day increase her mobility. In the meantime, she was unable to play and run like other little girls. Cari hoped that Cowboy Dreams could help Caitlin and enrolled her in September 2000.

The perfect horse for a little girl is a little horse and Taffy was that horse. A special bond developed between Caitlin and Taffy that started with their first ride. Caitlin loved the power and

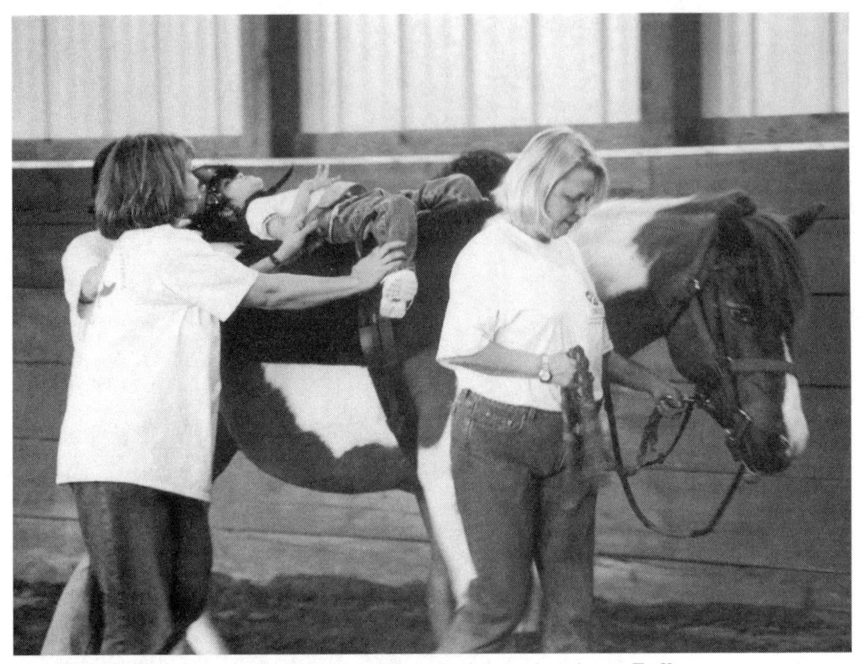

Volunteer walkers lead Caitlin as she stretches back on Taffy. (Photo by Cathy Lockhart)

freedom that riding Taffy gave her and Taffy loved Caitlin's gentle pats and kisses on her muzzle. Caitlin experienced a new relaxed and happy feeling that followed each hippotherapy session. It lasted all day and helped her sleep through the night. After Caitlin's first ride on Taffy, Cari noticed an immediate difference in her daughter's body. "When I carried her into the barn [for her first session, her legs were so tense that] she could hardly get her legs around my waist. After her hippotherapy session, her legs were so loose, she had no problem getting them around my waist." In time, other problems also diminished. If Caitlin sat facing forward on Taffy to steer (hippotherapy uses lots of positions), it didn't bother her so much to separate her legs. The typical painfulness of Caitlin's physical therapy for tight and spastic muscles virtually disappeared. In place of the hunched over little girl who sat miserably looking down and complaining of how tight she was, a tiny, giggling sprite was sitting up, pulling her shoulders back, and

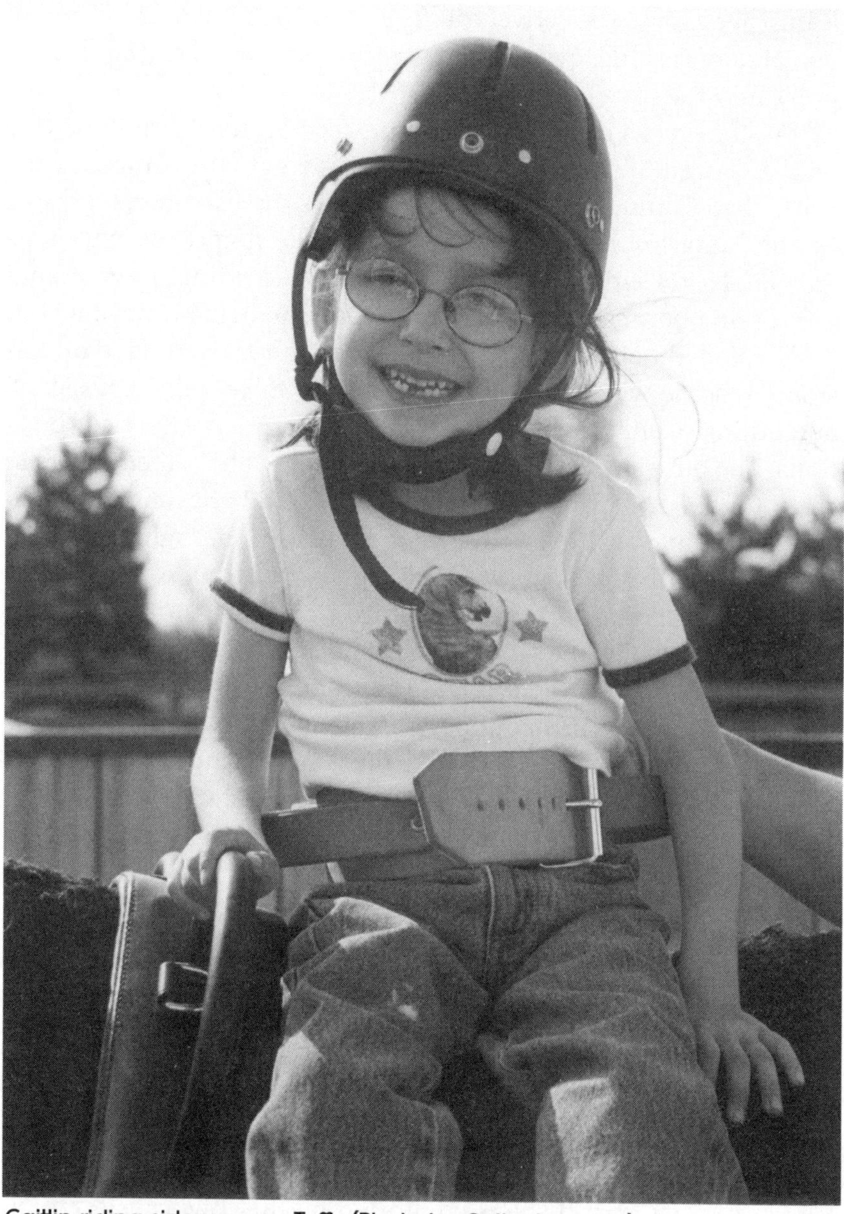

Caitlin riding sideways on Taffy. (Photo by Cathy Lockhart)

lifting her chin to see between Taffy's ears. Caitlin was in control of one area of her life, even if that control was only steering a small pony, and she loved it.

As the weeks of hippotherapy proceeded, new activities were added to Caitlin's therapy. Taffy accepted without complaint the tasks that Caitlin's therapists devised to increase her reaching: Caitlin hung toy rings over Taffy's ears; she tugged on her mane and tail. Next Caitlin lay on her stomach while riding. Taffy did not mind being bumped in the flanks; she gave no irritable head tosses. Taffy enjoyed her routine with Caitlin and may even have understood in some way that Caitlin needed her to be quiet, consistent, and strong. Caitlin enjoyed her rides on Taffy and the new positions. The once shy little girl was shy no more. She rallied others to cheer for her and made requests that they find her even more challenging riding positions. The new challenges brought new successes for Caitlin.

It has been almost two years since Caitlin began riding Taffy. Her confidence fills the air each time she is in the riding arena. The planned surgeries to help release hip and leg muscles that both doctors and therapists had thought were inevitable have been canceled. As long as she has Taffy, her hips and leg muscles are better than any surgery could make them. Before she started riding, Caitlin's long fatigued days led to sleepless nights, which led to much higher muscle tone, which led to more fatigue: an unending cycle. Because she sleeps better on the days she rides, her muscle tone is lower, more normal, the following day. She has more normal cycles of rest and alertness.

Before Taffy, Caitlin had no interest in, much less time for, hobbies. Her life was an endless cycle of therapies. Since Taffy, she envelops herself in horses. She reads about horses, colors pictures of horses, watches *any* movie with horses, and has real friends who also love horses. Her new friends are children she has met at Cowboy Dreams, and she is comforted by this. She calls them on the phone, plays with them at birthday parties, and talks with them endlessly about how much they love horses. She has a normal little girl's life and she has hope for her future.

The older girls in therapeutic riding do positions on their horses that she has not yet tried. They are role models for her. She watches them and wants to ride like they do. Caitlin is herself a role model for younger children with cerebral palsy. Their dreams grow because she gives them hope. With hippotherapy they all have a future that was not possible before.

As for Taffy, at twenty-some years, she is an old girl, even for therapeutic riding. But as long as she seems to enjoy it and as long as her body can handle it, she will patiently carry children on her painted back, her gentle nature, and her perfect size and gait making her the ideal therapeutic horse.

Chapter

Nine

Cookie

On an easel at the head of Thelma Gilkeson's casket in Jellison Chapel sat a photograph of Cookie and Thelma. Before Thelma died she had requested that it be at her memorial service. Its presence conveyed a significance beyond words, repeated in the many smiles and hand shakes that Jeanene got. In the pew, as Cookie lay beside Jeanene, he listened to the musical sounds that were the words and the purpose of the service. Cookie's requested presence at the service bespoke the strength of Thelma's attachment to him.

Thelma's daughter later wrote to Jeanene: ". . . and so my thoughts turn to someone so important to her and who gave her such pleasure and joy . . . [you and] Cookie . . . what you do is such a wonderful service . . . how many conversations we had about your visits, and how important they were to her. . . ."

Thelma loved the silky brown-and-white Brittany Spaniel with the moon in his eyes. Over nine years of visits a strong love had grown between them. Through times of laughter Cookie brought Thelma delight; in her sorrows, comfort. He was insistent in his

Thelma Gilkeson's hand and Cookie.
(Photo provided by the *Topeka Capital-Journal*)

attention. Nudging his muzzle under her good hand until it covered his silky crown, he would heave a deep and relaxed sigh. His head remained fixed there as long as Thelma and Jeanene conversed, his eyes following each speaker. Jeanene learned to take her place on the floor next to Cookie, looking up at her beloved teacher and friend.

When Jeanene first drew Cookie out of the animal control truck, she noticed there was something about his eyes. Despite his youth—he was only five months old—just looking into his eyes made her feel better.

Jeanene adopted Cookie, who developed Parvo—a merciless disease from which an untreated dog can die within days. Jeanene gave Cookie many hours of gentle companionship, stroking and brushing him, soothing him during his life-threatening illness. As his basic tranquility developed, Jeanene recognized that she had in her care the perfect dog for visiting the elderly and frail.

Cookie and Jeanene Hoover, 1996.
(Photo by Zercher Photo, Topeka, Kansas)

Thelma Gilkeson and Cookie. (Photo by Jeanene Hoover)

People living in nursing homes often have complex adjustments to make. They have lost their belongings, the familiarity of their life-long surroundings, their community of friends, and sometimes their families. The reasons for their moves usually include the loss of their ability to care for themselves, to do the things they used to do without thought. They often suffer severe bouts of intense loneliness and loss. In the face of all that, the simple friendliness and openness of an animal gives new residents some simple pleasure that most others, at home in their comfortable worlds, take for granted. In the midst of their loss, the responsiveness of an animal can bring a much-needed connection and soul-healing. That feeling of connectedness was there with Thelma and Cookie.

Thelma was by no means without people who loved her. Even before Jeanene and Cookie came into her life her devoted daughter, grown grandchildren, and great-grandchildren regularly came to see her. And she had her lifelong love of music. Her talent with the piano may have been struck down by a stroke, but still she enjoyed the pleasure of piano recordings, especially those of her talented daughter. Despite the stroke that made her left side a useless encumbrance, Thelma had a world rich in experiences to impart. She and Jeanene carried on long and intimate conversations, always with Cookie giving Thelma his full and concentrated attention. Jeanene showed her the same thoughtfulness. Detecting this similarity between them, Thelma once remarked to Jeanene that it was no wonder that Cookie was the way he was. He was a mirror of his owner—gentle, patient, and devoted.

So, what exactly did Cookie do that made him such an inspiration and earned him the love of Thelma and many others? When he first met Thelma he was the catalyst for conversation, a reason for Thelma and Jeanene to dig deeper into and explore the boundaries of their common interests. Over the waves of time that they shared, Cookie was physically and emotionally present for Thelma. Simply put, he was *there* for her.

Most of us know, when we have heartache or joy to share, if the person we share our events, our feelings with, is emotionally present. We recognize when that person cares about what is essential to us. Cookie did that. He did it exceptionally well. Why else would Thelma request that a dog be present at her memorial service? Even more than that, why else would she want that dog's picture displayed by her casket? When she needed someone to be emotionally present, Cookie was there.

Chapter Ten

Andy

Today, as she had done so many times before, Janet climbed the three steps of the concrete porch and knocked on the door of the small house shaded by graceful old trees. There was one step for each year they had known each other, she thought.

Sam opened the door and welcomed her inside; he led her once more into the clean and modestly furnished kitchen. Janet's memory of Sam's father, Philip, stretched before her like a winding road that grew tiny and indistinct in the distance. She could almost see him now. He had sat across from her in the metal-framed kitchen chair, with Andy's curly head poked up between his knees. That was how they connected, through touch. As she mentally continued down that road, she remembered how Philip always wanted to give her food. It was Philip's Italian heritage, Janet guessed, that compelled him to feed his guests. Janet never left without some small gift from Philip—a Pepsi, some tomatoes from his garden, a handful of peonies. Andy, Janet's hospice therapy dog, soon learned that Philip was a willing source of all kinds of exotic (to dogs) nuts. Pistachios, peanuts, cashews . . . Andy got them all. By the time they had finished their hospice visit, Andy and Philip had cleaned out the supply of nuts, and Andy fairly swayed down the porch steps. Once, gradually over the hour-long visit, Philip had fed Andy a whole can of cashews. Andy was in heaven with his nut fest, but Janet had watched the debacle with a prescient vision of her paunchy pooch, his upset stomach, and her having many extra washloads of towels. Janet was sure that, of the two of them, she would regret the indulgence more than Andy. The next time she had seen Philip, she had gently suggested that dog treats might be a better choice for Andy. The following week, on the Formica dinette table next to Philip, sat a very large bag of dog biscuits. Another hour later Andy again waddled out the door.

Now Sam sat in the same little chair next to the Formica table that his father had, recounting his father's history. How Philip had died at ninety-seven years of age. How he, Sam, had cared for his mom and dad for thirteen years. How, at ninety-four years, Philip had broken his leg and hip while cutting down a tree that had fallen on him. Imagine! thought Janet, ninety-four years old and still hardy enough to cut down a tree.

"After that," said Sam, "he couldn't walk anymore."

Sam didn't mention how his father had also suffered from congestive heart failure (CHF). Perhaps, mused Janet, he didn't know that Philip had congestive heart failure. As a nurse though, Janet knew that Philip's slackened blood pressure, his fluid-filled ankles and hands, and his complaints of never having enough energy signaled that his heart was failing.

Because Philip didn't show symptoms for a while, Janet's hospice care turned instead to the pain Philip suffered from the tree falling on him. At one point Philip's health rallied. "He did well enough," Janet said, "for me to discharge him from hospice. Then we came back." They came back because Philip was in end-stage congestive heart failure.

Philip brought his family to America from Sicily and Naples in 1924. He worked on his cousin's farm in St. Paul until he found a job with the railroad. When he wasn't working for the railroad, he

Janet Janes and Sam Alesso, Philip's son, along with therapy dogs Andy and Alice. (Photo by Donald W. Smith)

Philip Alesso and Andy in Philip's kitchen. (Photos by Sam Alesso)

was working in his garden, growing tomatoes and cucumbers and other vegetables, or working on his house.

Philip was not the kind of person who talked easily with others. For Philip, a hardy and self-sufficient man, words were mostly unnecessary. He showed his family his love by his actions. Burly Philip also loved animals. Over the years there were three family dogs, all of them poodles, laid to rest in the animal cemetery several blocks down the street. In his later years, Philip walked every day to first visit his wife's grave and next the pets'. It was because of this quietness and love for animals that Andy was the perfect hospice intermediary. Andy's job was to help Janet help Philip. He relieved Philip's day-to-day tedium and relaxed the accumulating layers of frustrating effort and pain.

Sam saw his father forget his cares and pain around Andy. He saw his father's mood lift when Philip knew that Janet and Andy were coming. Philip would ask Sam about their arrival and then order Sam to make ready. "Go get some treats," he would instruct Sam. When Janet and Andy were not around Sam saw a cheerlessness in his father. "His head would be hanging down when I came home . . . he went into himself. He was quiet."

Some of Andy's skills as a therapy dog have been naturally present from conception. His curly coat, for instance, is so plush

Janet Janes's therapy dog, Andy. (Photo by Donald W. Smith)

and soft, it's like falling into clouds. Touching him is especially soothing to hospice patients who normally suffer when touched, even lightly. His appearance is comical. Andy's eyes are buried so deeply in his curls and his coat is so thick, he looks like a walking Berber rug. Some of Andy's talents are most certainly a combination of genetics and upbringing. He is happy-go-lucky by nature. His silly antics can bring laughter into otherwise somber rooms. Laughter is a potent medicine for unrelenting pain and the grief of loss that accompanies terminal illness. Andy is capable of adapting his mood to fit the needs of the situation. He can calm his exuberance when the occasion calls for quiet. When there is no strength in someone, Andy can be a steady presence, sitting or laying quietly by the ill person's side. Genetically, he is predisposed to this work. His breed, the Portuguese Water Dog, is known for its endurance and deep spirit; these two traits make Andy a supreme companion in providing for the needs of those whose daily lives are overshadowed by the knowledge that they will soon die.

The testimony of Andy and Janet's gifts to Philip is found in Philip's kitchen. The picture on the refrigerator door is still there because Sam promised his dad that he would not take it down. It shows Philip, sitting in his kitchen chair, Andy wedged between his knees, both of them smiling.

Chapter

Eleven

Kodiak

For years Sharit Kelley had dreamed of combining her many and disparate passions. From her studies in wildlife management and her later work as a zookeeper, zoo educator, and national park ranger, she gained a deep appreciation for the capacity of animals for unconditional giving. The people who most needed this were those who had a physical, mental, or emotional handicap. Sharit returned to graduate school for a degree in Marriage and Family Therapy and training as a drama therapist. Theater offered people opportunities to safely explore who they are and who they can be. Sharit's passions for animals, people, and theatre naturally culminated in her involvement with the Barrier Free Theatre Troupe using animal-assisted drama therapy for people with disabilities.

Sharit needed a well-trained animal, preferably a dog. She applied to be a puppy foster-parent through Canine Assistance, Rehabilitation, Education and Services (C.A.R.E.S.) of Concordia, Kansas, and within months was holding her new charge: a curly-tailed Swedish Elkhound pup rescued from the local animal shelter.

Sharit knew that C.A.R.E.S. went through a careful evaluation process with each puppy they placed, so the puppy had already shown that he was smart, a people dog that was comfortable with changes, and not easily frightened. Kodiak was all of that and more.

Kodiak quickly learned how to behave around people, sitting mutely by the door, for instance, when someone came to the house, or waiting by Sharit's side until he was given the "ok" to visit with someone. The training he received was more intensive than for many therapy dogs. He was trained in basic obedience and socialized with both people and animals for goal-directed therapy work training, which usually takes at least one year. Through C.A.R.E.S., Kodiak was also trained as an assistance dog. Assistance dogs are trained an additional six to twelve months to prepare them for working with the specific needs of a disabled person or group, under the guidance of another trainer-owner. Their training is similar to that given service dogs, trained to eventually be owned by and to assist one person with disabilities.

At eight months of age, young for any dog, Kodiak was sufficiently mature and experienced that he could begin his official therapy visits. These were done through the local therapy pet organization, Pets-n-People. Kodiak learned to be around all kinds of people and circumstances and, in short order, was certified as a professional therapy dog. In the course of his visits he learned to accept unintentionally tight hugs or pats that were really just clumsy thumps. He grew accustomed to wheelchairs and other special equipment.

Kodiak could sense the level of energy in a room and match it. He was patient and careful with people who were hesitant to approach him. With "at risk" youth in the park who were excited and full of energy, Kodiak bounded and chased with exuberance every thrown ball. With elderly people at assisted living facilities, Kodiak mirrored the quiet, reserved demeanor of his elderly admirers. He would calmly devote himself to one person, spending as much time with that person as required, before he moved on.

In the Barrier Free Theatre Troupe, Kodiak's sixty-seven pounds of compact, muscular, and agile energy was sometimes intimidating. If people were not used to being around dogs or just a bit hesitant around this one, they would initially hang back from Kodiak, watching, waving, or blowing kisses to him from across the stage. What they discovered was reassuring: a friendly and well-mannered dog that did not come near if they did not want him to.

The ultimate confirmation that Kodiak was not only friendly but also a potential friend was when the troupe members learned that he knew several hand signals. What a delightful revelation that was! Troupe members were often unable to communicate effectively with other community members—the unfamiliar listener had difficulty understanding their words or their sign language—but here in front of them was a curly-tailed, laughing-faced dog who understood at least fifty words! His ability to recognize their signing gave each person in the troupe an unexpected sense of community, belonging, and independence. Sharit saw Kodiak gain trust with troupe members. She encouraged troupe members to praise Kodiak and give him rewards for being responsive. The result, in her eyes, was a growing sense of empowerment within these people who controlled so little of their lives. Troupe members became friendlier, not only with Kodiak, but also with each other; they showed more confidence in themselves and were more willing to communicate with fellow cast members.

What Sharit most enjoys about the Barrier Free Theatre Troupe, as does everyone, are the performances put on at the close of the season. Everyone in the troupe, including Kodiak, participates in the process, from the initial story conception to the final curtain call. Members design sets (Kodiak provides encouragement), create costumes (Kodiak models or sports his own), choreograph dances (Kodiak learned a line-dance along with the rest of the cast), and think up dialogue and songs (Kodiak sings, too). Everyone has a role; they direct and they rehearse. They advertise and promote ticket sales. When the play is finally presented to the public, they are part of the enthusiastic audience. The most recent play of the Barrier Free Theatre Troupe had a cowboy theme, with Kodiak cast as "Coyote Yak" and Judy Inglesbe as one of his partners.

Sharit (*left*), Kodiak, and Judy. (Photo by Kevin Kelley)

Judy Faye Inglesbe is fifty-five and wheelchair bound. As a result of cerebral palsy her muscles are now locked in many places and her speech is limited to a handful of words, pronounced only with great effort. Before she knew Kodiak those words were limited to a few names (her sister, Joyce, is "Oose") and similar monosyllabic words. Judy had always been a happy person with an infectious laugh. But the difficulties of "doing for herself" combined with others' willingness to do too much for her, created a more passive person.

Judy's mom, Cecile, had watched Judy interact with Kodiak during the months that the play was developing and saw her daughter emerge for the first time from habitual passivity. Judy, she saw, was noticeably more independent. She indicated her needs and was more insistent about them, rather than accepting what was given to her or looking to caregivers for guidance. She was thinking for herself and communicating her thoughts more

Sharit (*left*), Judy, and Kodiak. (Photo by Kevin Kelley)

Sharit (*left*), Kodiak, and Judy. (Photo by Kevin Kelley)

effectively. Her memory, too, was clearer. Judy was noticeably more involved with her world when Kodiak was around. When the time came for Judy and Kodiak to perform for the public, they took center stage without hesitation.

They sat side by side on the stage, heads thrown back, cowboy hats secured below their chins, bellowing with gusto their cowboy song. The words were indistinct, but that didn't matter. Everyone enjoyed watching the faux cowboys belt out their lusty tune. There was no doubt that Judy was having the time of her life. Her canine partner's carefree abandonment in his own "singing"—a ranging, off-key dog yodel—doubled her pleasure. They completed their song and sat radiant in the spotlight. Judy's life and the lives of the Barrier Free Theatre Troupe were rich that day, and Sharit saw her dream become reality with the help of her animal assistant, Kodiak.

Chapter

Twelve

Sparky

Laura Fertitta was baffled by the request. Her three-year-old daughter's therapists wanted her to begin to ride. Laura thought that Chessie's motor problems (poor strength, balance, muscle tone, and motor apraxia) could possibly get better, but she didn't quite understand how riding a horse was going to help. And were they also telling her that horse riding could improve Chessie's speech? Laura was skeptical. But nothing else was working. What did she have to lose except a little time and money?

When Francesca Madalena Fertitta was born Laura knew she was perfect, but as time went on, she seemed to develop slowly. By the time Chessie was eighteen months of age Laura had stopped reassuring herself that her daughter was only getting a slow start. Chessie could not sit up by herself. When she was propped up and reached for a toy, the rest of her body inevitably followed her hand until her head thumped on the floor. And words! Laura had looked forward to the day that Chessie would say "Dada," the first word of every child she knew. It was unsettling to Laura that Chessie only squealed and shrieked her enthusiasm or crossness.

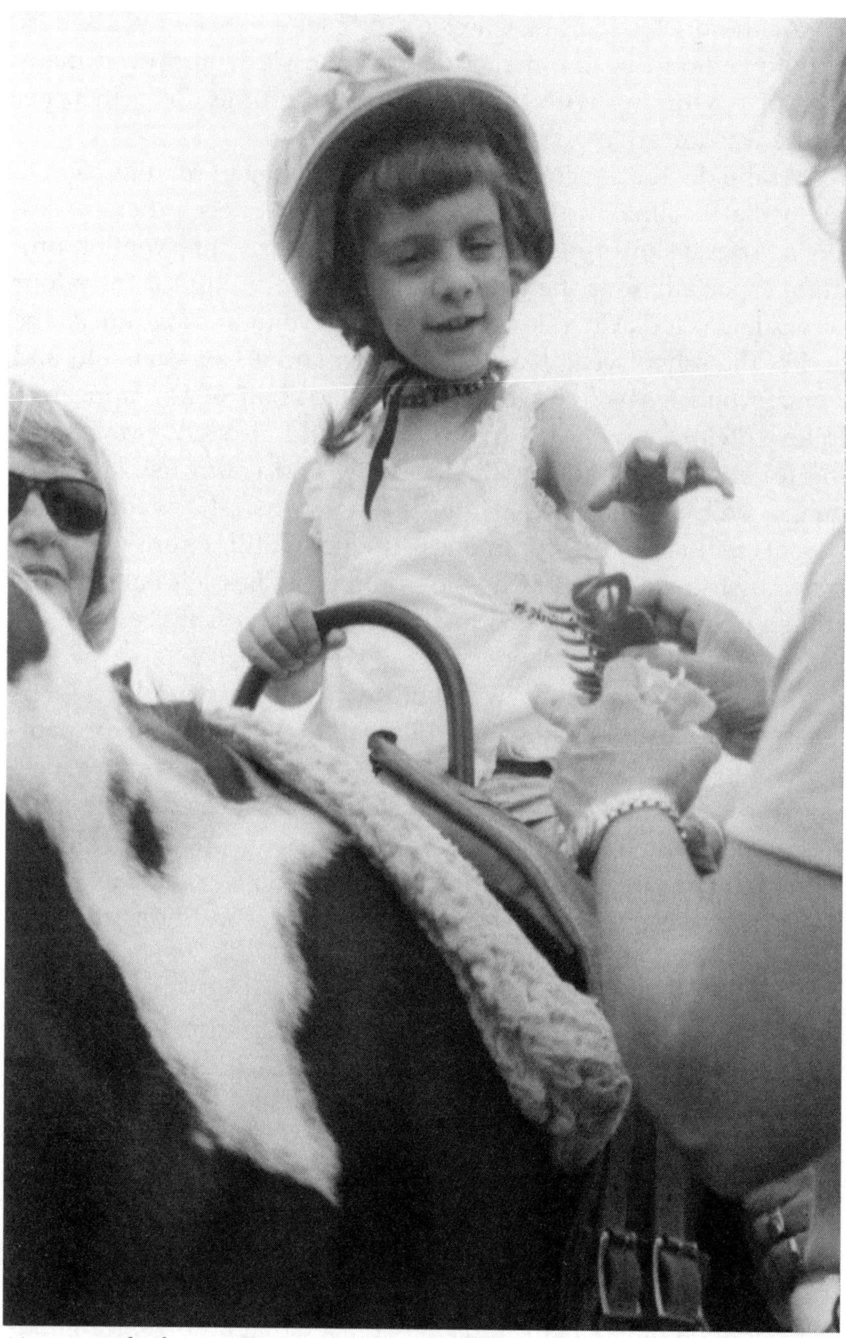

Miss Linda (*left*), and Chessie riding on Sparky. (Photo by Nicholas McIntosh)

At the doctor's office Chessie's problems were given names: hypotonia and developmental delays. Chessie would need many therapists working with her for a very long time. Even then her future was uncertain.

At the doctor's recommendation, Laura began regular trips to the pediatric clinic for Chessie's physical and speech therapy sessions. After a short time she switched to an early intervention program especially for infants and children. Every day for forty-four weeks Laura took Chessie to the early intervention program. At the end of the school year, Chessie was two and a half years old and could proudly say five words: "Da" (dad) had finally appeared, followed closely by "Ma" (mom). She could also say "eat," "ore" (more) and "uce" (juice). Chessie knew and could use sign language well enough so that others knew what she wanted. Her squeals and shrieking became less frequent. Still, Laura was frustrated by the slowness of Chessie's progress. Chessie's balance was still so poor that Laura had to sit behind when Chessie rode the little merry-go-round at the mall.

The next doctor visit was with a pediatric neurologist. His assessment was that Chessie had verbal apraxia—the motor pathways of the brain were not sufficiently developed. The brain's message for motor movement was not clearly transmitted to the muscles of the mouth and tongue, so words are not pronounced clearly. Although she had no idea what verbal apraxia was, at least Laura knew she wasn't crazy. There was a real condition affecting Chessie. Now Laura could move forward.

For two more months Laura continued Chessie's speech and physical therapy sessions. At the close of each session Laura would meet with Chessie's therapists to talk about Chessie's needs, the meaning of the apraxia diagnosis, and what else could be tried to help her daughter One of Chessie's therapists proposed trying hippotherapy.

For Laura, it had been a year and a half of diligently scheduling, driving to, attending, and returning from multiple therapy sessions. She was less than enthusiastic about the "horse thing." In fact, she remembers being very pessimistic that a horse, for goodness sake, could make her daughter talk. "But," she explained, "I was willing to give it a try." She paused before adding that, actually, she was *sure* that it wouldn't help. She set out to prove it.

Chessie's first hippotherapy sessions were scheduled at the Muscular Dystrophy Council for Special Equestrians with her physical therapist, volunteer walkers at her sides, and a horse leader. Laura watched, armed with a healthy dose of uncertainty. Chessie, Laura cautioned, had always been frightened around large animals; it was best if people didn't expect too much. But Francesca Magdelena Fertitta immediately proved her mother wrong. Not only did she ride the small horse named "Little Bit," but she fell in love with riding. During the car ride home and all through the next week Laura listened to her tiny daughter pretend she was riding. The house echoed with Chessie chirping,"Oh Bi!" ("Go Bit!"). Laura was only partially mollified.

After two therapeutic riding sessions, Chessie again went on the merry-go-round at the mall. She no longer needed her mother to hold her.

In a standard operant experiment, the effectiveness of an experimental technique is verified by first applying it in a situation, removing it, and then reapplying it in the same situation. If the expected results occur the first time, stop when the technique is stopped, and recur when the technique is reapplied, the technique is highly likely to have caused the result. In effect, this is what happened with Chessie when the location of her hippotherapy program changed.

The initial test phase had been with Chessie's first two hippotherapy sessions, after which she was independently balancing on the merry-go-round. Between the time the program ceased to be offered at one site and then was restarted at another, Chessie necessarily missed some therapeutic riding sessions. This was the second phase of the "experiment." Laura recalls what happened:

"It was during this down time that we went to the mall again, and this time [Chessie] could not ride the carousel by herself. She was always asking 'Er Bi?' ('Where's Bit?')." She also started to revert to squeals and shrieks.

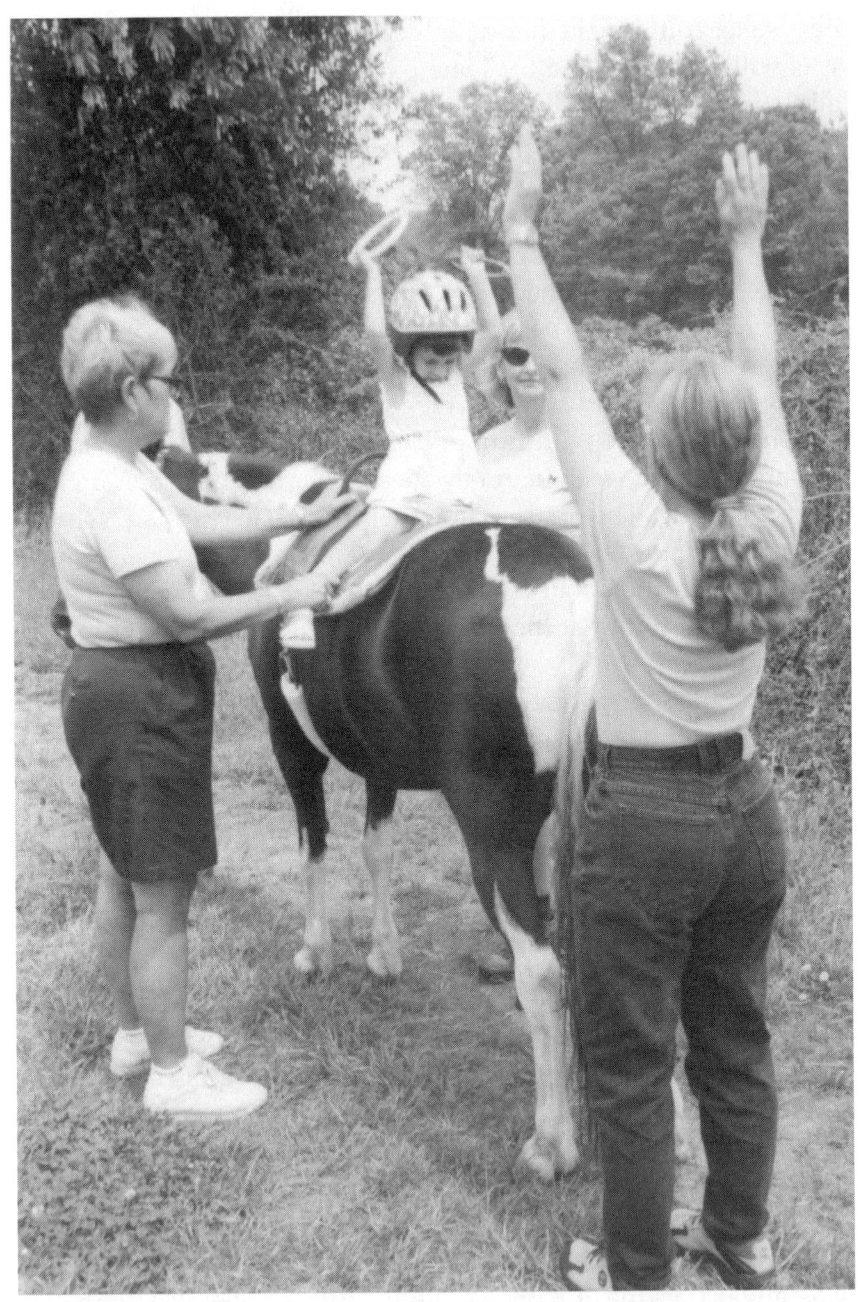

Miss Jean, Chessie with rings on Sparky, Miss Linda, and Robin Korotki (*back facing camera*) during hippotherapy session. (Photo by Nicholas McIntosh)

Chessie on Sparky with Miss Linda. (Photo by Nicholas McIntosh)

Sidewalker Miss Linda, and Chessie on Sparky. (Photo by Nicholas McIntosh)

The retest phase was, of course, when Chessie started riding again, this time on a new painted pony. "In September," said Laura, "Francesca met her new horse, Sparky. She was so excited! She got on and said 'Go!' It must have been love at first sight."

As Chessie's regular therapeutic riding sessions resumed, she could again sit by herself on the merry-go-round. The effectiveness of hippotherapy was validated. Now Laura was convinced.

Sparky is a gentle, ten-year-old black-and-white painted pony that easily tolerates the toys, sounds, and unusual rider positions that are typical of therapeutic riding. His ability to accept the unusual with aplomb created in Chessie the security to eagerly try new things. On top of Sparky, Chessie's motor and speech skills continued on an accelerated track.

Before Chessie started hippotherapy, she could not blow out her birthday candles, balance while sitting or standing, reach across her midline or say any intelligible words. After less than a year of hippotherapy, Chessie can play nerf basketball on either her left or right side and can walk and sit with confidence.

The success that most impresses Laura, though, is how Chessie's talking improved as a result of therapeutic horse-riding. In six months, Chessie went from saying "Go Ar-E" to clearly enunciating, "Go Sparky." Now, Laura says, anyone can understand the words to the songs that Chessie sings nonstop: "Happy Birthday," "Twinkle Twinkle Little Star," and Barney's "I Love You" song.

Laura stated, "I kind of get the 'I told you so' impression [from the therapist]. And I have *never* been so glad to have been proven wrong!"

Laura was not the only one impressed. Chessie's doctors were amazed and even went so far as to say that Chessie is far better off than any other child they have seen with her type of problem. Even the experienced physical therapist is impressed with Chessie's progress. "Today," she said, "Chessie even amazed me. She read the letters in her name and then told me that it spelled 'Francesca.'"

Chapter

Thirteen

Jerry

"There is a quiet comfort and feeling of calm that envelops me while holding Jerry," wrote Debi. "When I look at him sleeping in my arms, sprawled out, tummy to the world, in the most vulnerable position possible, I'm more easily able to show my vulnerabilities, knowing that I, too, can be comforted." This was Debi Demiglio's explanation of what happens when she has therapy with Diana Lee and Jerry, a seven-pound Brussels Griffon and certified therapy dog.

Debi was practical about her decision to start therapy with Diana. She had not been to an individual therapist before, although she had been in a therapy group for women from codependent relationships. The group work had been long and intense and helped her move out of an abusive relationship. Debi now wanted to concentrate in therapy on her career choices.

When she started seeing Diana, Debi worked in a job she detested; she did not understand why she continued. She wanted to explore her future career options as well as the reasons for her

Diana Lee and therapy dog Jerry, 2002. (Photo by Karen A. Pomerinke)

staying stuck in a job she did not like. Diana helped her work through the feelings she clutched tight about herself. "Diana brings out the best in me," said Debi. "She sees me for who I am, not who I thought I was." Diana was talented before she included her muscular squirt of a dog in her practice. Including Jerry allowed her to expand upon her talents.

Diana treats clients for depression, anxiety, grief, post traumatic stress disorder, addiction, recovery issues, and abuse. In Diana's estimation, Jerry, or Dr. Jerry as he is sometimes addressed, provides an environment of safety, security, humor, and especially comfort to her clients. He sits in a person's lap, looks intently at someone, and, if there is a need, licks the tears off their cheeks. Besides being a source of comfort, Jerry's snorts and snores are often good for breaking the tension that sometimes builds in therapy. Depending on your point of view, his timing is either incredibly good or incredibly bad but the results are the same: he gives people an opportunity to momentarily step away from their personal catastrophes and relax. Diana believes that Jerry has a gift for putting people at ease. Research certainly bears this out—petting a dog has reliably been shown to lower the blood pressure of the person stroking the dog—and the therapeutic result for Diana is that clients more readily talk about issues and more quickly and efficiently move through therapy.

Debi began examining her career before Diana started to include Jerry in her practice. But Jerry's presence later made a significant difference for Debi. There were unexamined events in her past that continued to cause her distress. For these more difficult issues, where trust of her own judgment was central to her improvement, having Jerry present calmed her. The feeling of emotional safety he gave her allowed her to broach tender topics, like the sexual abuse she had suffered as a child. "I needed to forgive [her abuser]," she admitted over the phone. "I thought I could do that. I thought that was all. That person cropped up in my life again as I was working on my career goals." Pragmatic career planning and forward movement in therapy collided head-on with Debi's past. Now she had to forgive an abuser who was again a part of her world.

When people are overcoming psychological pain there is a strong urge to "let sleeping dogs lie." Without Dr. Jerry, Debi may not have had the courage to begin laying to rest the pain from her past and to forgive this person who had so carelessly taken away her sense of safety. Jerry "made me calm instead of reactive," said Debi. "I was able to hear [the idea that forgiveness] is a process."

Like Debi, many of Diana Lee's clients were in therapy with her prior to the introduction of Dr. Jerry. Having watched many of them before and after Jerry was, she says, enlightening. "Without a doubt they can more readily access the grief, trauma, and [painful] memories with him here in my office."

Diana can be this effective by herself. But working together with her therapy dog, Jerry, her clients work more easily through their issues. The clients that Diana sees stroke a dog, relax, and talk.

Chapter

Fourteen

Tegan

If you have visited in the northern suburbs of Chicago, perhaps at a local hospital, nursing home, or a home for people with developmental disabilities, if you have been in a county facility for drug-addicted and abused women and their children, or had a hospice nurse in your own North Chicago home, you might have seen Fred with his ten-year-old Irish Setter, Tegan. You would certainly have noticed how Tegan's deep red coat reflects ripples of light as he moves. You might not, at first, have taken in his gait, how its cadence is different from what you would have expected. Pad-pad-pad—pad-pad-pad. If the rhythm struck you, you might have stopped and watched him more closely. Then you would recognize that Tegan is missing his right front leg and shoulder. He had bone cancer when he was three years old.

Doctors gave Tegan an 8 to 10 percent chance to survive the cancer. Even then they did not expect him to live longer than two more years. Tegan beat the odds. He was a therapy dog before his cancer; and seven years after his cancer and the loss of his leg, he remains a therapy dog.

Tegan. (Photo by Marilyn Putz)

For people like Ingrid, confined to a wheelchair at Westmoreland Nursing Home, Tegan's visits were the difference between waiting to die and continuing to live. Ingrid, an elderly Swedish woman, had no surviving family when Tegan began visiting her about five years ago. In the beginning Ingrid was hesitant about touching Tegan and preferred instead to watch him visiting with others. Her fears were tempered by the repeated gentle greetings (and satisfaction) Tegan gave others, and one day Ingrid asked Fred if he could bring Tegan closer to her chair. She extended her hand toward Tegan and found a cool, silky head pushing under her fingers; the small weight that Tegan felt signaled him to turn his head ever so slightly and slowly stroke his long pink tongue across her hand. For Ingrid, the tactile experience stirred memories of other loving touches, of infants once nestled contentedly in her arms. All this was communicated between Ingrid and Tegan in an instant.

Fred and Tegan's visits to Westmoreland now included conversations with Ingrid. She loved the joy Tegan exuded at her touch, the remarkable devotion that Tegan showed her. Fred and Tegan were a respite from the dreariness and lonelieness that were common features of her day-to-day life. The visits, the relaxation and comfort that she felt from Tegan's smooth fur, his joyful presence, and his endlessly rocking tail made her days happier and

Marilyn, Fred, and Tegan visit with some nursing home residents. (Photo by Marilyn Putz)

gave her something to look forward to. As Ingrid's worsening eyesight and increasing pain complicated her ability to move, she still delighted in Tegan's visits. Although she often felt separated from the world in which she lived, the big red dog brought her a sense of connectedness and purpose.

Eventually, Ingrid's health declined to continual pain and near blindness. She stopped waiting in the hall for Tegan. Tegan, though, knew that Ingrid was missing and searched her out. He nosed through the hallways until he found the place where her scent was strongest, then entered that room. Gazing around, Tegan spotted Ingrid sitting in her chair and bounded toward her, his tail wagging a mile a minute. Now beside her, Tegan rested his head on her lap and gazed upward into her face. Upon realizing that "her" red dog was there, Ingrid's face immediately lit up and her body relaxed. For the few moments that Tegan was by her side, Ingrid forgot her afflictions. She beamed at Tegan, chatted with Fred, and was at peace in the knowledge that Tegan truly cared for her.

Even when she was confined to her bed and any movement

Marilyn and Fred with Tegan. (Photo by Marilyn Putz)

was difficult for her, she would still reach out to touch that soft and reassuring head. The nursing home staff and doctors would gather around and watch in wonder and satisfaction at the differences in Ingrid when Tegan was around. Somehow, Tegan knew that this kindly but very tired woman needed the cheer that only he could offer. Tegan gave Ingrid his all.

Though Ingrid passed away, "we would venture a guess," said Marilyn, "that [Ingrid] is somewhere, waiting to become reacquainted with her adoring friend, Tegan."

People in the northern Chicago suburbs who watch the big red dog still do not quickly discern his physical uniqueness. What they see is strength, command, and assurance of life in his movement. With Tegan beside them, people sense that there are some things worth living for on the other side of the pain. His very presence conveys to others, "There is hope."

Fred Putz with Ingrid and Tegan. (Photo by Marilyn Putz)

Chapter
Fifteen

Magic

This is the story of a little girl who started in therapeutic riding and blossomed into an accomplished dressage horsewoman. This is also a story of a group of people with a shared commitment to therapeutic riding. Like a close-knit family, they share activities, care about each other's progress, and do whatever it takes to help each other succeed.

In the waiting area of the indoor arena, on wooden stools around a rough-hewn table, I sat with some of these people and talked. Susie Bailiff, the director of Mounted Eagles, reminisced about its start.

> SUSIE: I never realized all the hoops I'd have to jump through to do it. . . . I nurtured the idea . . . for many years before it became a reality. . . . [Mounted Eagles] started after reading an article in *Reader's Digest* about two autistic boys . . . the experiences those kids had . . . the seed got planted.
>
> [I]t was nine or ten years before the program became a reality. . . . [A]ll of a sudden, one year, it was like the doors flew

> open and it was time, and everything fell into place and everything went smooth, and we started receiving funds from different organizations.

Mounted Eagles chose Isaiah 40:31 to illustrate their commitment and their faith: "But those who hope in the Lord will renew their strength. They will soar like eagles; they will run and not grow weary, they will walk and not grow faint." This was, for Susie, the heartbeat of all that Mounted Eagles had been and would be. She worked tirelessly, but after nine or ten years there still was little to show. When Susie saw an eagle glide along the roofline of the barn on her property, right toward her at eye level, it seemed like a sign. Within a year, in 1990, Mounted Eagles Therapeutic Riding was born. Amy Roth was one of the first to enroll.

When Amy was born ten weeks prematurely to Diane and Paul Roth, she was diagnosed with hyaline membrane disease, which progressed to bronchiopulmonary dysplasia and asthma. She had spastic diplegia cerebral palsy and seizures. Diane, who had been studying to be an occupational therapist, saw that her daughter's future depended on immediate and massive interventions. Diane started researching treatments for spasticity, a disorder of the muscles where muscle groups tighten and will not release. The studies Diane found convinced her that horses, and only horses, stimulate exactly the right muscle groups and nerves that Amy would need for walking.

> DIANE: I learned that a horse's walking pattern is the closest to mimic a person's walking pattern. When a person walks, their pelvis moves in three different dimensions at the same time . . . a horse's walk does this, too.

Diane started looking for places for Amy to ride. But Amy was so tiny and fragile—her respiratory status was touch and go until she was nearly nine—that riding was sporadic at first. It was not until she was four years old that Amy could consistently attend her therapeutic riding lessons with Mounted Eagles.

Amy leads Magic out of the barn. (Photo by Donald W. Smith)

Four-year-old Amy grooms her horse. Amy's mother, Diane Roth, is at the left. (Photo by Paul Roth)

> AMY: My first memories of riding in Mounted Eagles are of a large outdoor arena, and of my mom taking my hand and leading me up the mounting block. The horse I rode was a gray gelding named Raffles. He couldn't have been very tall, but to me he looked enormous. I remember how scared I felt that first time, as I was led at a slow walk around the arena.

Amy found reassurance with the people on either side and at the front leading her, talking to her. She took to riding like a porpoise takes to jumping.

> AMY: Eventually I forgot about my fear as I began talking to the side walkers who held me on Raffle's back, their hands on my legs. We'd do certain exercises on the horses, like "flying," where we'd stick our arms out to the side like an airplane, and then do twists and turns. Then there was ball toss, where we'd line up two riders side by side and toss a large ball to each other. It was a good thing the horses were so calm, because we usually missed each other and hit the horse in the head or chest. Also, there was the animal game, where one rider would ride past the other and

make an animal sound, and the other rider would guess what sound it was. Mine was too easy to guess, because I did a horse every time.

Because Minnesota winters are too cold for riding outdoors, Mounted Eagles had to close for the season during its first two years. The facility did not have an indoor arena and could not offer Amy the year-round lessons that she needed. With no therapeutic riding available, Amy's ability to stretch her muscles degenerated toward her pre-riding levels.

Over the next two years, a pattern emerged. Each spring Amy would have to regain her former achievements. Through summer and fall she would strengthen and build her muscle use. Each winter, with the closing of Mounted Eagles, she would regress. If she was going to make any significant gains, Amy needed a year-round riding program. Eventually, Mounted Eagles moved to the Clare B. Training Center, a location with an indoor arena. Amy's lessons were now headed by Toni Wasilensky, a gifted dressage teacher and trainer.

TONI: I remember Paul . . . carrying her in; she was this adorable little girl with great big chocolate brown eyes. . . . I thought, Oh my goodness. She looks like a china doll.

Amy's development surged. The little china doll, so quiet and reserved, loved riding and everything that went with it. She also, it turned out, had an impish side.

AMY: After we rode, the Mounted Eagles crew would have us groom the horses. I loved that part. Raffles would stand so quietly as I tried to use the brush. It was like he didn't care what we did with him. Mom would stand nearby and watch. She usually had a cup of coffee in her hand, and I would purposely try to flick as much horsehair as I could into her cup, just because I loved the annoyed tone in her voice when she'd say "Amy, stop that!"

Amy on Tolono. (Photo by Diane Roth)

By the winter of 1996 Amy's muscle tone was relaxed enough that she could finally bend over, dress herself, and tie her own shoes. For the Roth family, as well as for Amy, these were major accomplishments.

That Christmas held other surprises for the Roth family. Nine-year-old Amy was the baby of the family of four; her only sibling was her college-age brother. When he returned home for the semester break, Amy and her brother resumed their good-natured teasing, which included her brother prodding Amy to try harder, to not be a baby. When he noticed that his little sister stopped using alternating steps up and down the stairs, he told his parents they shouldn't let her get away with this. Amy protested that she was trying as hard as she could. Over the next few days Amy's walking down the steps stopped altogether. She didn't complain that her body hurt; she just adapted. Amy began the butt-slide method of going up and coming down the steps. Her parents and her brother now investigated.

DIANE: Amy never had very good strength on her right side. I didn't notice how she was starting to scoot around until [her brother] confronted us. [He] thought Amy was babied, that we needed to push her to stand up. Amy said her legs weren't strong enough and in five or six days it got really bad. She couldn't walk. We took her to Gillette's Hospital.

Diane and Paul found out that Amy had a spinal cord infection called transverse myelitis, a rare and acute neurological condition with severe inflammation of the spinal cord. As a result of the infection Amy could no longer use her legs. The doctor quickly put her on antibiotics and ordered physical therapy three times a week. Amy responded to the treatments. There remained some residual weakness in her legs but within a month Amy was again able to walk and the scary specter of paralysis was kept at bay.

Amy continued her traditional therapies and in the spring restarted her therapeutic riding lessons. Her posture improved, her fine motor control got even better, and she was able to hold her own balance. Amy was ready for more strenuous goals.

Mounted Eagles staff and families believe that involving each child in setting their own riding goal develops self-esteem.

For her first goal Amy chose riding without side walkers.

DIANE: Her balance was so bad when she first started, plus the [lack of] strength in her arms and hands, the thought of maybe her horse spooking and taking off on her just scared me so bad. . . . Amy set goals that were so high. Not just as a mom, but as a *therapist,* I just thought they were not realistic. I did not want to break her heart or discourage her.

In a matter of a few months Amy quietly and purposefully achieved her first goal. She took it all in stride, but not without taking advantage of the opportunity to playfully tease her mother.

DIANE: [S]he met that goal and when she met it she rubbed my nose in it.

Amy riding in the mountains of Colorado. (Photo by Diane Roth)

Amy's second goal was trotting, which required greater control of her body. Though many doubted that little Amy was ready for trotting, Amy was ready to show everyone that she could do what she set out to.

> AMY: My favorite thing to do was trotting. I had two people, one on each side who would run beside me while somebody led the horse. The only way they could get me to do it was to promise that if I fell, one of them would throw themselves underneath me. We only trotted for a couple of strides, but by the time we walked again, I was laughing.

No side walkers, then trotting, led to several other goals and accomplishments. Amy was the star of Mounted Eagles. She was riding as well as, and often better than, any rider who had normal muscle function. She had made friends and earned respect and admiration at Mounted Eagles across the years. Besides the deep sense of family, there seemed little else for therapeutic riding to give her. After much deep thought and emotional upheaval, Amy chose to leave therapeutic riding.

> AMY: Sam [Amy's best friend] and I did our first Mounted Eagles horse show together. I quit Mounted Eagles about a year after that show, having found that my riding level had gone beyond what the program could teach me. That was probably one of the hardest decisions I'll ever have to make. Being out there every week, I grew so attached to the instructors and the horses. They were—and still are—a second family to me. I knew I wasn't really LEAVING leaving because I'd still see them on Thursday nights. I'd still be out at the barn to volunteer. But at the same time, it wouldn't be me up on the Mounted Eagles horses. It'd be some other kid riding in my spot. In some ways it just didn't seem fair.

Although Amy had quit therapeutic riding, she had no intention of leaving riding altogether. Leaving Mounted Eagles was rather a step toward unassisted riding. Her natural choice, for many reasons, was dressage riding with private lessons. Like most

Amy astride Magic during a dressage riding lesson, 2002.
(Photo by Donald W. Smith)

Amy gives Magic a much-loved ear massage.
(Photo by Donald W. Smith)

people, Amy started dressage with her horse on one end of a lead rope and her teacher, Toni, on the other.

With dressage, Diane hoped, Amy would have something on which she could focus her ambitions. She was taken off guard when Amy upped the ante and said she had one more challenge for herself (and for her mother's courage). Amy wanted to go trail riding in Colorado.

> DIANE: It's one thing to be riding inside the barn in an enclosed arena but, it's like, [on trails] how do you stop a runaway horse when it's spooked?

Amy, earnest and single-minded, earned her trail riding trip by developing control of her horse. She and her parents drove to Colorado. Diane learned that Amy, in fact, knew her own abilities very well and was willing to risk failure to experience the challenge. Diane could trust Amy to make decisions for herself that were based on self-knowledge, self-trust, and a willingness to work hard to reach her goals.

TONI: She's an excellent example of being able to do anything you truly want to do, if you just work at it. With the limitations that she seems to have physically, they really don't limit her. I mean, she just finds a way to get around it and she does it anyway. What an incentive for the rest of us, you know?

Amy finally got her own horse about two years ago. She paid for the gray gelding, Magic, with money she had saved and through the generosity of people, sometimes unknown, who wanted to support her accomplishments. Amy is now working in dressage riding to use her balance and muscle strength to control her horse. Only as a last resort is she allowed to use the reins. In June of 2002, Amy competed in her first open dressage competition.

She won first place.

Chapter

Sixteen

Lucy

In the few months that Chandler (Chan) Rudd and I have corresponded, it has became clear to me that Chan is a remarkable person. His Golden Retriever, Lucy, is equally remarkable. Together the impact they make is both lifelong and life-giving. Chan will tell you that whatever they have accomplished in their volunteer activities has been because of Lucy. "What happened to me happened to everybody who ever met [Lucy], and that is I fell in love with her." Lucy is affectionate with everyone and revels in their attention.

For the last three years Chan, his wife, Dee, and their two Goldens, Lucy and Bennie, have made weekly visits to the local hospital and nursing homes, and facilities significantly more distant, to share with people who are recovering or suffering. On any given week, there are fifteen to twenty people they see on the rehabilitation unit of Exeter Healthcare. These people, Chan remarked, are often pleasantly surprised that a hospital would allow dogs on the unit. During these visits patients are diverted

Bennie on Chan Rudd's boat. (Photo by Chandler Rudd)

from their pain and loneliness and receive the tender care of a friendly person and their loving and well-trained animal.

It was in this spirit of giving that Chan, acting on the request of a mutual friend, contacted Rebecca Hardy's mother, Nanette. He asked if she would like Chan and his wife to bring their therapy dogs to visit nine-year-old Rebecca (Becca) at the Barbara Bush Children's Hospital in Portland, Maine. This was where Becca would soon receive another round of chemotherapy for her bone cancer, formally known as Ewing's Sarcoma. Rebecca would be there six days, Nanette told him; it was her third-to-last round of chemo and the family was optimistic that Rebecca would continue to show progress. Nanette's fearless daughter was on the upswing of her battle with cancer and would be delighted to "play with a dog" in her hospital room.

Chan paved the way for their visit with hospital staff. When he and Dee arrived with Lucy and Bennie at the hospital's main entrance, the volunteer at the information desk, unaware of any therapy animal visits, commanded, "Hold it just a minute with

Lucy. (Photo by Chandler Rudd)

those dogs!" Chan stopped short. Just where, she grilled Chan, did they expect to go with those dogs? Chan responded mildly that they were expected on the sixth floor to visit a nine-year-old girl named Becca. But the volunteer would not have any of that; this man was trying to bring *dogs* into the hospital!

"You can't just walk up there. I have to call up there and make sure it's okay. What's her last name," she demanded.

Chan could feel the heat rising in his face as he groped for the memory of Becca's last name.

"Uh . . . I uh, I don't remember," was all he could stammer.

The elderly woman standing before him seemed to grow taller. He could imagine her in a sergeant's uniform and smacking a nightstick against her open palm.

"I did call and talk to the charge nurse and she said it was fine," Chan said, glancing up and then sideways.

"What," Chan heard The Uniform slowly say, her eyes narrowing at The Spy Who Came in with the Dogs, "is *her* last name?"

Chan playing with Lucy. Bennie sits at the right. (Photo by Dee Rudd)

"Uh . . . I don't remember that either," Chan muttered. He was starting to feel that things weren't looking too good for him.

Irritated at his impertinence, the Officer in Charge of Everyone Who Comes in the Door made her judgment. "I'm going to call and see if someone knows you are coming."

While she dialed a number, hidden from Chan's view, he began wishing he could quietly sneak into the waiting elevator. If there had been a way to escape gracefully, he most certainly would have taken it, but the commitment he had made to the little girl on the sixth floor kept him glued to his place in the lobby, waiting for the sergeant major of the hospital volunteer corps to finish her phone call.

After she finished conferring with someone in charge of the Children's Oncology Unit, she handed the phone to Chan and told him to state his case again.

"My wife and I have come from New Hampshire to bring our two therapy dogs to visit Becca. She's in Room 641."

Lucy in her cart at Goldstock. (Photo by Chandler Rudd)

"Oh!" responded the voice. "You have therapy dogs! Come on up!"

Chan and his troops headed for the elevator. The dogs, having never felt the rise of an elevator before, sniffed the floor in curiosity and looked up at Chan and Dee, who just laughed at them. They exited the elevator to find a more welcoming response from the staff of the Children's Oncology Unit.

When they entered her room, Becca was lying in bed, doodling on a pad of paper, having just undergone another round of chemotherapy. Becca sat up in alert interest when she saw five-year-old Lucy being wheeled into her hospital room.

As a five-week-old puppy, Lucy had been found in a town dumpster, left there by someone who did not want a dog with birth defects so severe that her whole back end was paralyzed. The little Golden was found by a woman who was familiar with "Yankee Golden Rescue," an organization that places Golden Retrievers in adoptive homes. She called Chan, who is on the organization's board of directors and has been affiliated with YGR for many years.

Lucy in her favorite place–the water (Photo by Chandler Rudd)

Would they take the little dog with the frozen back legs and the big eyes? They would.

Lucy moved to southeast New Hampshire, where she went through months of veterinary assessments and treatments, including water therapy (to see if she could regain function in her back legs) and surgery (to remove the one back leg that hung awkwardly out to her side). Lucy eventually came to live with Chan and Dee while they trained her for adoption. In the process of watching Lucy's efforts and seeing Lucy react with sadness when she had the occasional disability-related bowel accident, Chan fell in love with the young, golden dog.

"If you had seen her when she was five weeks old—you know those cartoon characters with the great big eyes?—well, she had those big eyes, and nobody could do it, nobody could put her down." Chan knew that the little dog was an extraordinary gift and decided to give Lucy a permanent home.

The door to Becca's room was closed when Chan and Dee arrived. Dee knocked softly and a small voice said "Come in!" The first thing Chan saw was Becca's sparkling smile, then her eyes, bright and warm. The first thing Becca saw was the cart that carried Lucy into her room. The cart was Lucy's mode of transportation for long trips. Becca watched in fascination as Lucy rolled in, instantly recognizing that Lucy, like her, had survived some mighty hard curve balls. After Chan lifted Lucy up and placed her next to Becca's tiny legs, Lucy arched her head backward until her big brown eyes were gazing into Becca's hazel eyes. Becca responded by stroking the soft fur on Lucy's neck, while Lucy continued to look up at Becca, as if she were her hero. Not to be outdone, Bennie placed his front paws on the bed at Becca's side and refused to move until Becca spent at least a little time stroking his fur. Chan and Dee showed Becca how Lucy walks, relying on her front legs for muscles and using her one hind leg like a kickstand, arching her back and flipping her hind leg up. They told Becca about Lucy's life before she came to be a part of their family and how she had won their hearts. They gave Becca a photo album filled with pictures of their dogs but intended also for when she finished her treatment and got the dog that her mother had promised her. They also brought Becca a plush Golden Retriever stuffed toy.

Many children are inspired by Lucy's gentle strength and affection. Some of the children face life-altering or terminal illness and trauma. For them, Lucy is an inspiration of hope. According to Chan, Lucy turns their perspective from feeling sorry for themselves to recognizing that, although difficult, their own trials are not insurmountable.

Chapter

Seventeen

Jeb

Jeb and Deb were the intervention of last resort for three-year-old Matthew Smith. He has autism. He was barricaded behind a communication wall that prevented him from processing language and kept him emotionally detached from others. A person observing him would see a little boy randomly bouncing from one location to another, like a metal ball in a pinball machine. Matthew played only by himself; he repeated in a rote fashion some activity or series of actions and never made eye contact with anyone, although he might momentarily glance in someone's direction. Only on objects did his gaze rest steadily—perhaps on a button, a shoe, an earring, or his own hand. He did not say words, although he could sound out repetitive, meaningless syllables. His special education teacher, Dana Romary, had studied autism and knew the common interventions. He had talked with the experts and had tried the suggestions they offered. So far, Matthew had not made progress. Dana and the rest of Matthew's preschool special education team were stuck.

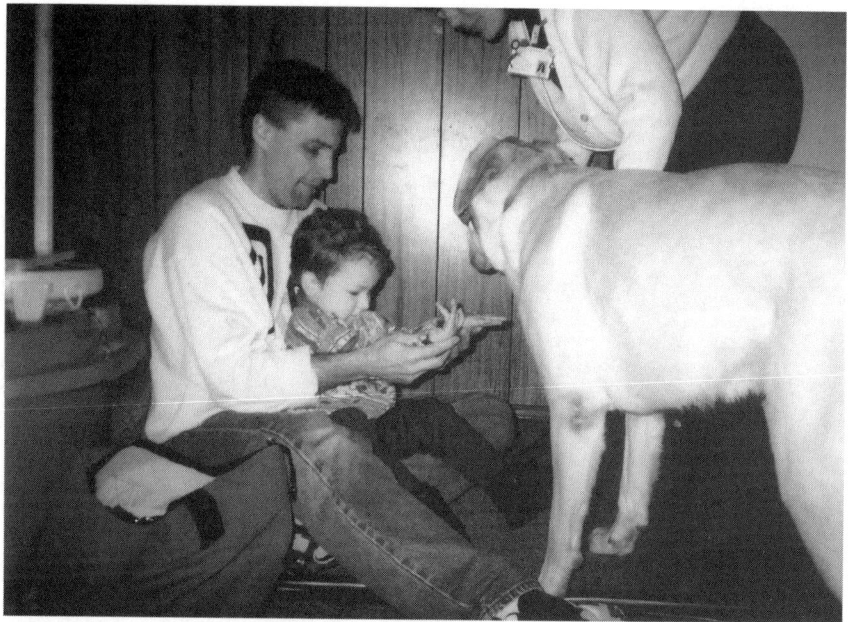

"Open your hand, Matthew." From left, Matthew's teacher, Dana, Matthew, Jeb, and Deb. (Photo courtesy of Deb Sellers)

Dana and one of the classroom therapists began looking for novel ways to connect with the little boy. One afternoon, the conversation eventually wound around to Deb Sellers and her new yellow Lab crossbreed, Jeb. Jeb had come to live with Deb six weeks before, as a certified assistance dog from Canine Companions for Independence. Might Jeb be able to break into Matthew's world? Dana called Deb and together they discussed this option. Working with Jeb was still new territory for Deb but there seemed little to lose. She consented. Dana then talked to Matt's mother, Melinda; she, too, agreed that it was worth a try.

Within a week four hopefuls were standing in front of Matthew's home. Melinda's group of munchkins surged out of the door and around the new arrivals, reforming into a larger group, that rambled back inside the house. Matthew's face was not among them. Once in their living room, Deb gave Jeb a "sit" command and he sat, whereupon he was immediately surrounded by little heads and little hands patting, rubbing, petting, and pointing on all

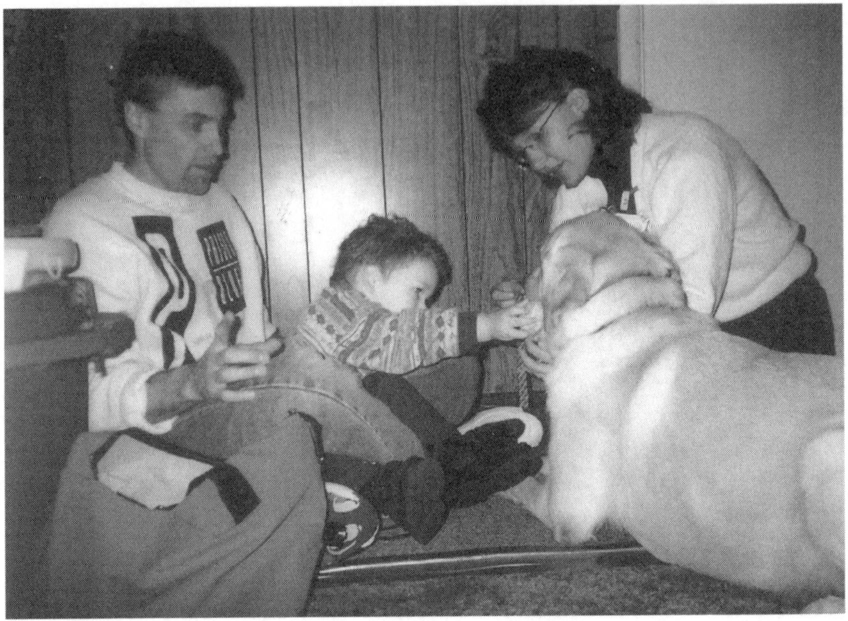

Dana holds Matthew as he, for the first time in his life, looks in the eyes of another being, Jeb. Deb Sellers is beside Jeb. (Photo courtesy of Deb Sellers)

parts of him, in all directions, all at the same time. Jeb sat very still, just as he had been directed, occasionally searching for Deb's face, looking for reassurance that he was not being abducted by aliens or in some other mischief. Deb, happy with her well-behaved partner, gave Jeb a smile and Jeb relaxed into the pleasure of the moment.

Once all of Matthew's siblings had had a chance to pet and talk to Jeb. Melinda and her visitors discussed where to work with Matt. Everyone agreed that Matthew would be most likely to focus under quiet circumstances. They decided that the best place would probably be the basement; they could shoosh away any children who were down there and it would be relatively quiet.

Matthew, having managed to walk with his mother down the stairs, took advantage of the opportunities the big, open basement offered. He climbed and ran and jumped and sat and paused for a second or two, then started all over again. No one in the room had worked with a dog *and* a three-year-old tornado before; their

Deb Sellers wtih Matthew and Jeb. (Photo courtesy of Deb Sellers)

unspoken hopes that Matt would zero in on the novelty of Jeb's presence were soon flattened. At eighty pounds, Jeb was hard to miss, but Matthew was oblivious to his, and everyone else's, presence.

During one of Matthew's flying passes through the room Dana scooped him into his arms; Matthew did not resist the physical contact. Dana sunk heavily to the floor with Matthew in his lap and Deb directed Jeb to "visit" Matthew. Matthew may have tracked Jeb in his peripheral vision as Jeb arose, crossed the room, and rested his head in Matthew's lap. Whether he did or not, he did not acknowledge any awareness of Jeb's presence. Despite Jeb's immediacy, despite the seven pounds or more of large dog head that came to rest on Matthew's crossed legs, Matthew continued to randomly look about the basement. It was good that Matthew was not backing away from Jeb and that he was tolerating the close proximity of two living, breathing creatures. Dana moved his large hands over Matthew's small wrists and hands. Together they stroked Jeb's head and ears.

A minute of this dog petting was, for Matthew, a very long time.

Before the little boy decided to change the activity, Deb decided for him. She took Jeb's ball out of her equipment bag and called Jeb to her. Jeb took the ball in his mouth and, as Deb instructed, dropped it in Matthew's lap. Dana again moved his hands around Matthew's wrists and hands. The two of them then grasped and threw the ball to Jeb. Although they repeated the game of fetch several times, Matthew showed no interest. He began to wriggle in Dana's lap. It seemed the experiment had reached its limits. Jeb dropped the ball in Matthew's lap one final time. The ball sat there in Matthew's lap while he fidgeted, not moving away from Dana but not showing an interest in the ball.

Then, after several seconds and by himself, Matthew grasped the ball in his two small hands. He crouched, half in and half out of Dana's lap. He looked at Jeb, extended his arms, then handed the ball back to Jeb.

There was an absolute hush in the basement. Not one of the five adults could believe what they were seeing. The photograph on page 127 captures this moment. It shows Matthew, reaching toward the big yellow dog, looking right into Jeb's face. The reaching and the looking were both small and huge things. Matthew had never been known to make eye contact. He had not interacted with another person or creature in his whole life, even for two seconds. A bit of magic hung in the air; each person felt they had scaled a summit of success. When Matthew finished handing the ball to Jeb, he went back to bouncing all around the basement, playing first with one toy, then another. The adults in the room began telling each other what they had seen, even though they were all watching the same little miracle and telling the same tale. It was as if they needed to speak the event into reality. Jeb lay down to rest.

The scenario went round and round in Deb's head as she packed her belongings. The other adults continued talking excitedly, engaged with each other instead of Matthew. Matthew was no longer the center of attention as he slowly walked toward Jeb, a toy held in both hands. No one paid attention to him until he plopped down next to Jeb. The adults stopped talking.

Matthew as he continues to interact with Jeb, a first-in-his-life event. (Photo by Deb Sellers)

Except for Matthew and Jeb, it looked like a room full of statues in Melinda's basement, everyone's eyes again glued on Matthew as he climbed over Jeb's back, rested his sharp little elbows in Jeb's side and continued to play with his toy. Deb wondered if Matthew was not so much interacting with Jeb as he was using Jeb as a piece of furniture. Just in case, she instructed Jeb to "stay." Matthew stayed, too, propped on Jeb's side, playing with his toy. Jeb could not have been very comfortable with the sharp little elbows in his side but he did not get up.

Matthew had already done two things he had never done before: look at and interact with Jeb. The pièce de résistance followed when, for a second or two, Matthew gazed right into Jeb's eyes. Matthew looked as if he were realizing there was more to this thing than something to lean on; this was a dog! Melinda grinned broadly, confirming for Deb what she had suspected, that Matthew had connected with Jeb, even for just a moment.

Deb's emotions remained at a peak long after both she and

Matthew went on to something else. She and her cohorts finished packing up their equipment, congratulated each other on their successes that day, and said their goodbyes to Melissa, Matthew, and the other kids. The group was still amazed that Matthew had made eye contact, had kept his attention on Jeb for longer than they had ever seen and, especially, that he had made an emotional connection with Jeb. Each of these were enormous steps for Matthew. That they occurred together, in one forty-five-minute session, was truly magical.

In telling this story for me, Deb reflected on its meaning: "A living, breathing, animal can not be replicated by any other modality. . . . We can observe the effects of animal-assisted therapy as we did with Matthew, but the question of why this worked when so many other things hadn't is not to be answered at this time. Perhaps animals truly do carry magic within them as a part of their nature.

Chapter Eighteen

Bleu

How could a large male cat overcome a long history of being abused and still like people? What sense does a cat have about safety and forgiveness when he has had his right ear shredded, buckshot blasted into his left hindquarters, and his left hind leg broken and left to mend crookedly? Would you or I return to any person, any place, that reminded us of that kind of torture? It is perhaps a bit of faith in people that allowed Bleu to bury his head in someone's neck, wrap his front legs around that someone's shoulders and purr. Who is luckier, Bleu or the person who receives Bleu's affection?

The people at the animal shelter guessed that Bleu was about three years old when they rescued him from an abusive owner. He had survived who knows how many life-threatening experiences. One family saw Bleu at the shelter and tried adopting him. But he cried all night and they returned him the following morning.

In the same community another cat, a well-loved cat named Wendell, had just died. Kim Shinn was broken-hearted, and Nan,

Angus Maitland at home with his cat. (Family photo)

his wife, knew that Kim needed the sit-in-your-lap, rub-against-your-leg, and nuzzle-your-neck kind of love that a cat gives to help him recover from losing Wendell. She went to the shelter, heard Bleu's story, and went home to tell Kim about the cat that had been taken out of harm's way. He would be euthanized before the end of the month if no one adopted him. Kim and Nan went to the shelter. "As soon as I took Bleu in my arms," said Kim, "he buried his head in my throat and put his front arms around my neck. It was pretty much love at first sight." Bleu, the big, blue-gray tabby with the floppy shredded ear, was now a member of the Shinn family, along with their Samoyed therapy dog, Sasha.

Bleu adapted to the weekly visits at Hadlow Hospice with aplomb and grace. Nan took him there whenever someone requested him. As Nan was preparing to leave from one of these routine visits, a middle-aged man she did not know approached her. His father-in-law, he said, was a patient here: Mr. Angus Maitland. Mr. Maitland had loved cats his entire life and was now near death. His family feared that Mr. Maitland, who recently had slipped into a coma, would soon die. Would Nan please bring Bleu to see him? Nan followed Mr. Maitland's son-in-law to the room.

Angus Maitland had neither spoken nor moved for several days. He lay still and silent in his bed, his breath shallow but regular. His family sat around him in the hard hospital chairs, watching and wondering. How much longer did he have? Would they have the chance to say goodbye?

Nan entered the room softly. She placed Bleu on the bed by Mr. Maitland's side and stood back. Mr. Maitland's wife and daughter-in-law moved Bleu to Mr. Maitland's chest, took his hands, and placed them on Bleu's head and back. "A kitty has come to visit you," they told him. They moved Mr. Maitland's hands across Bleu's back and then let go.

Slowly and so slightly that someone watching could not be sure, Mr. Maitland moved his hand across Bleu's head and back. Then more definitely, he stroked the blue-gray fur. The big cat's gentle purr filled the room. The family was stunned. They had thought Mr. Maitland's hands would never move again.

Mr. Maitland continued stroking Bleu over the next five minutes and his hand came to rest, finally, on top of Bleu's back.

Bleu at home. (Photo by Nan Shinn)

Nan Shinn and Bleu outside Hadlow Hospice Center. (Photo by Kim Shinn)

Watching this, Nan asked Mr. Maitland if he was tiring. Mr. Maitland shook his head, "No." Mr. Maitland started stroking Bleu once more. The family quietly voiced their astonishment. When Mr. Maitland's hand once more came to rest, Nan, as she does with every patient she visits, asked if he would like Bleu to visit again sometime. Mr. Maitland mouthed a quiet, "Yes."

Over the next three weeks, Nan and Bleu visited Mr. Maitland five more times. Each time Mr. Maitland petted and stroked Bleu; each time when Nan asked if he wanted Bleu to return, he nodded or spoke his affirmation. It was clear that Mr. Maitland found pleasure in Bleu's presence, but his strokes across Bleu's back were visibly weaker with each visit that Nan and Bleu made.

Mr. Maitland died a little over three weeks after Nan and Bleu first came to see him and his family. His family was prepared for, even at peace with, his death. They felt they had been given a gift: Bleu had shown them that the person they loved was still present although he lay so still. That gift was irreplaceable.

Bleu has brought peace and pleasure to other hospice patients.

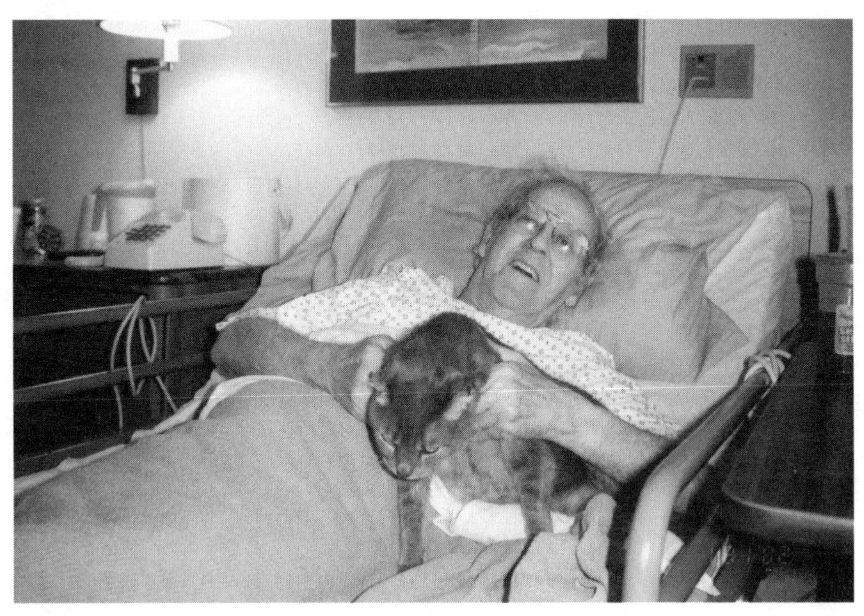

Frank Gill with Bleu. (Photo by Nan Shinn)

Frank Gill grew to love Bleu so much that he requested a photo of Bleu be placed in his casket. Bleu's successes gave Nan the idea to bring their second cat, a five-month-old, frisky male named Baldwin. Baldwin became a favorite of Michael LaChica, a forty-four-year-old man with Amyotrophic Lateral Sclerosis (also known as ALS or Lou Gehrig's disease). When Baldwin came to visit Michael, Nan and Michael's mother, Ruth, would often shut the door to Michael's room to let Baldwin run and jump and climb. Baldwin's antics brought Michael respite from the progressive paralysis that characterized his disease. Baldwin, because of his kitten status, was considered an intern at hospice, but interns can show considerable insight. At about six months of age, Baldwin sat in Michael's lap for forty-five minutes. Anyone who has owned a kitten knows what a rare thing this is.

In 2000 Bleu was honored as Jacksonville's Cat of the Year. He took the honor with catlike dignity.

Chapter

Nineteen

Timmy

On August 4, 2000, Scooter Ross was diagnosed with an inoperable brain tumor, a brainstem glioma. Most children do not survive more than seven months after being diagnosed with this type of cancer. As of January 2003 Scooter had lived with his cancer for thirty months; he is the first child to survive this long. His spirits, and those of his family, have ebbed and flowed around the course of his illness, the necessary radiation and chemotherapy treatments, the two-and-a-half-hour (one-way) trips to Dallas, Texas, and the isolation from family and friends. The flow of support has been nourished by the dedication and care of the volunteers at Ronald McDonald House and Linda Smith's three-and-a-half-year-old Rottweiler, Champion Ironwoods Primetime (Timmy, for short), a dog who wears bunny ears and angel wings. Scooter's family believes that Timmy is the reason that Scooter's spirits have remained so high. The mutual love of boy and dog started during therapy dog visits at the Ronald McDonald House and grew through doggy e-mails and telephone messages.

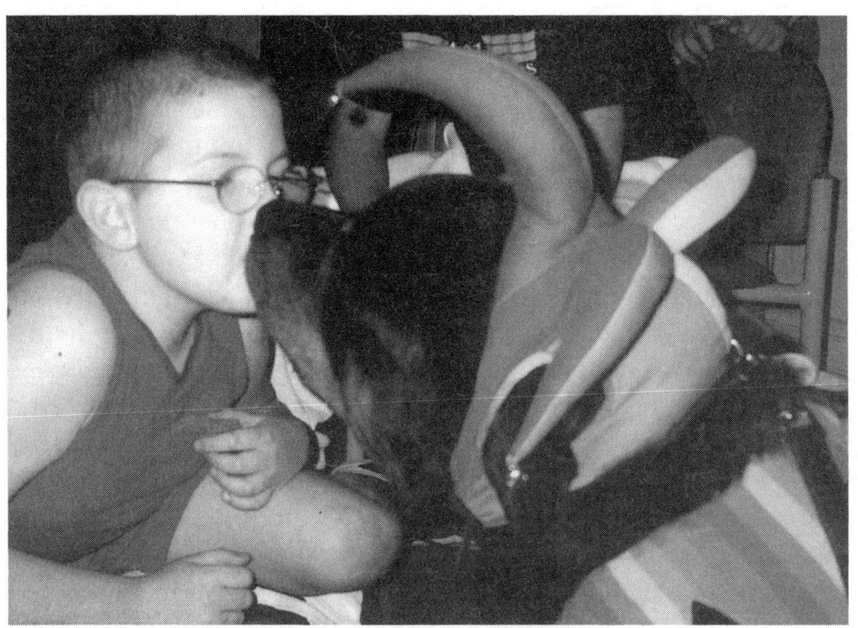

Scooter with Timmy as a jester. (Photo by Linda Smith)

Ten days after being diagnosed and during their first stay at the Dallas Ronald McDonald House, Scooter and his mother, Michelle, met Linda and Timmy. Timmy was one of several dogs visiting that Tuesday and, as she usually does, Linda had dressed Timmy in one of the seven costumes she has made for him. Linda knew that she needed to minimize the scareyness that Rottweilers are associated with, and help people to see that underneath Timmy's imposing size beats the heart of a bunny rabbit. In fact, that is one of his costumes and it is pretty hard to think that a dog is threatening when he is dressed up as a bunny rabbit.

That Tuesday Scooter saw a large, muscular, black-and-brown dog wearing a cowboy hat, a neckerchief, a vest with an American flag stuck in the side, and flowered sunglasses. Timmy was anything but intimidating.

Scooter and Michelle made weekly trips to Dallas for his cancer treatments while his father and big sister stayed home. The family had never been separated before. Scooter missed his family and he especially missed his eight-month-old yellow Lab puppy. To help him cope with this loss, Michelle and Scooter wrote down the days that the therapy dogs were supposed to visit at the Ronald McDonald House and scheduled their weekly Dallas trips to coincide with them. In particular, Scooter liked Timmy. Scooter and Michelle would wait in the main floor meeting area for the dogs to arrive and Scooter would watch the door, waiting for Timmy to walk through. As soon as he saw Timmy, Scooter would call his name. Timmy would zero in on Scooter's voice and search until he found him; the two would play together the whole visit. Michelle found that Scooter was happier after Timmy's visit, that he handled better the difficulties of chemo treatments and the isolation of their trips when Timmy was around. Timmy and Scooter's relationship grew, not only during the Tuesday visits, but in between. Linda made sure that there were extra, private visits between Scooter and Timmy. While they were separated, Linda and Timmy sent e-mails and left telephone "bark" messages on Scooter's home answering machine.

From: Scooter121989
To: Rotdawgs
Hey Timmy,
Just thought I would tell you goodnight
Love,
Scooter

From: Rot dawgs
To: Scooter121989
Hi Scooter,
This is yur Buddy Timmy xo. . . .

I'm sending YOU a night-night wet, slobbery, drooling kiss too!

Hey ~ Tanks Rotts for the good night mail. (My Mom got mad when I read your mail about the purty girls and drooled all over the screen and key board!)

Can you keep a secret?

My Mom doesn't know I'm on her computer again. Hey! I even learned how to use her credit card on the web! Wait til the one hundred pounds of doggy cookies arrives in the mail. . . . Ya think she'll know it was ME that ordered them?

Scooter and Timmy the cowdog. (Photo by Linda Smith)

Next: I need to learn how to drive. That way I can borrow the van and come up there and see my Best Buddy, then we could go cruising down Main Street and howl and drool at the purty gals.

(Do ya think we could sneak into Hooters and not be noticed?)

I can't wait for your burthday party at Hooters again!

Talk to ya tomorrow My Bestest Friend.

Rotts of Cookie Kisses and Slobbers,

TIMMY

X O

There were times when the chemotherapy treatments were especially devastating. Scooter became weak and listless; his normal appetite was replaced with nausea and vomiting. During these times Timmy's play with Scooter had a different quality to it. He was gentler, more tentative. He hung back, waiting for Scooter to initiate the play, and sat quietly by Scooter's side when no play came.

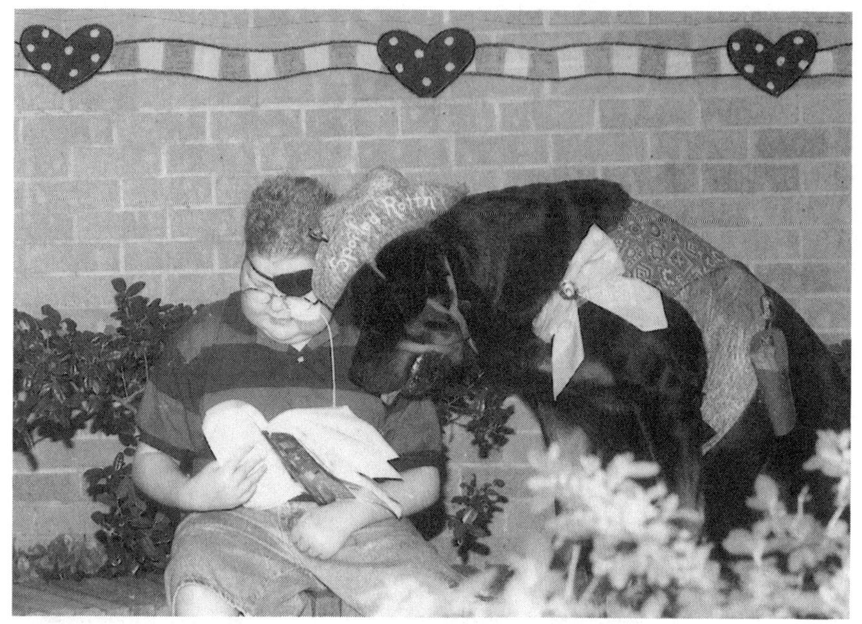

Scooter at Ronald McDonald House, reading to Timmy. (Photo by Dawn Robinson)

Scooter's chemotherapy is, according to Michelle, a trial-and-error process of trying to find the chemo cocktail that will stabilize his tumor. Since no child has lived as long as Scooter, his doctor says that Scooter is writing the book on treatment of brainstem gliomas. In a roller-coaster fashion, Scooter and his family never know when a chemotherapy trial stops working. Timmy does.

One Ronald McDonald House visit occurred on the same day that Scooter had tests done to see if he was still in remission. Toward the end of the visit Scooter left the meeting area to lay down in his room; Timmy followed Scooter's scent. Finding his room, Timmy lay down in front of Scooter's door. When it was time to go Linda came to get him, calling him to "come." Timmy refused to move. Linda had never seen her normally obedient dog refuse her commands in this way. She had to physically drag Timmy away. On the way home, Timmy was miserable; he could not be cheered. He remained unresponsive for the next two days. Timmy refused to leave his dog crate (an open pen) and refused to

Timmy in his bunny rabbit costume. (Photo by Linda Smith)

eat. When Linda called the director of the Ronald McDonald House and heard that Scooter's tumor had doubled in size, she put two and two together. Timmy, Linda knew, was reacting to Scooter's returning cancer.

Perhaps this was a coincidence; Timmy could have had an upset stomach or a short-term bug. But several months later, when Scooter again was no longer in remission and his tumor had tripled in size, Timmy responded in very similar ways. Linda and Michelle believe that it is Timmy's love for the little boy that leads him to be so keenly sensitive to Scooter's illness.

From: Scooter121989

To: Rot dawgs

Hey, Timmy.

Guess what. I went and saw some trucks that looked like monsters. They were really big. Well, just wanted to say hi to my bestest buddy. Wish you could have been there with me.

Love always,

Scooter

I miss you.

Postscript: Scooter's battle with cancer ended in January of 2003.

Chapter Twenty

Isabelle

Anne Tomasko and Isabelle, her certified St. Bernard therapy dog, were making a requested hospital visit to an eight-year-old girl who had refused to talk to anyone for four days, not staff, not friends, and not family. This little girl, we'll call her Sara, lay curled and facing the wall, indifferent to anyone who came into her room. Her mother was getting desperate for someone, anyone, to break through her daughter's wall of silence. When Anne and Isabelle entered the room Sara lay with her back to the door, as she had been all day. Sara's mother urged her to turn over.

"Look, Sara! Someone special named Isabelle is here to visit you!"

Sara remained fixed. Isabelle, determined to say hello, walked over to the bed and placed her giant head on Sara's pillow. Isabelle began snorting and snuffing loudly at the back of Sara's head. Sara impassively watched the wall, but a after a few minutes of this, Sara's curiosity about these strange noises overtook her. She slowly turned away from the wall to rest her gaze on Isabelle's liquid

Isabelle, certified therapy dog. (Photo by Anne Tomasko)

brown eyes and soft brown-and-white head. Sara, astonished, jumped up on the bed and screeched, then hurled her arms around Isabelle's neck, holding her tightly.

Sara's mother watched in amazement, distinctly aware of what lay abandoned at the foot of Sara's bed: It was Sara's favorite toy, a large, stuffed St. Bernard.

In the corridor, hospital staff heard the commotion and swarmed toward what sounded like an emergency in Sara's room. Nurses rushed in, only to find that Isabelle had so soothed Sara that both lay quietly on the hospital room floor, Sara's arms still wrapped tightly around Isabelle's neck.

Sara's doctor was called and soon joined the amazed throng of hospital staff at Sara's door, looking intently in at the intertwined child and giant dog. The doctor firmly wanted Anne and Isabelle to stay at the hospital forever. Barring that, he wanted them to stay the day.

Isabelle and Anne did stay with Sara and her mother for quite some time that day. And just after they left Sara finally broke her silence, talking about her wonderful new-found friend, Isabelle.

A soft, stuffed dog can be a balm to a child's unbidden terrors. Sara's stuffed toy had ceased to provide her comfort, leaving her mute and alone. Only when the toy suddenly seemed to come to life, in Isabelle, did Sara feel safe again and able to talk.

Chapter

Twenty-One

Missie

"Lorn," said Tracey, his physical therapist, "was a drug addict."

At eleven years of age Lorn had already survived encephalitis, or inflammation of the brain. His teen years were undoubtedly difficult. He would have had learning problems, perhaps needed special education classes, and, almost certainly, endured the derision of other adolescents who labeled him a "dummy" or "retard." In his twenties, maybe before, Lorn began using drugs. His chronic drug use caused three cerebral vascular accidents (strokes) that left Lorn with muscle weaknesses down his left side and around his mouth. His speech was slurred and he had difficulties swallowing, memory problems, poor impulse control, and poor judgment. He could no longer live independently. After he left hospitalization he lived in supervised facilities in Texas, then moved nearer to his family, to the state facility in North Dakota formerly known as the Grafton State School. Here Lorn has lived, worked, and received therapy for three years. His needs include physical and occupa-

From left: Carole, Missie, and Lorn. (Photo by Donald W. Smith)

tional therapy to improve his left side weaknesses and speech therapy to address his oral motor and speech problems. Cognitive activities have been of primary importance; without improvement in his attending to tasks and without a decrease in his impulsivity, it was unlikely that Lorn would reach his goals of living and working in the surrounding community.

In his first two years at the Developmental Center Lorn peppered his daily routine with refusals. If his job became difficult or boring, he quit. If his therapy sessions required more effort than he was used to, he would suddenly disappear. Even when he was trying, Lorn's attention span was usually only a few seconds long; his supervisor or therapist would remind him to continue many, many times. Lorn was pretty much stuck, making little or no improvement in his therapies and progress toward his goals. Then Carole and Missie began volunteering at the North Dakota Developmental Center.

Carole Torgerson lives in a small community near Grafton, North Dakota. Grafton is by no means a large community—less than 14,000 people—and the adjacent town where Carole lives is even smaller. There are no resources for learning about animal-assisted therapy in either place and none within driving distance. But Carole is resourceful and has enough energy and ingenuity for a whole community. Carole used the Internet to buy books on animal-assisted therapy; she found online organizations that supported animal-assisted therapy and through them enrolled in AAT courses. She sought out the help of people in distant communities and followed their expert suggestions, and she drove and flew to the areas where AAT certification was done. In record time Carole and her Golden Retriever, Missie, became a certified Animal-Assisted Activity/Animal-Assited Therapy (AAA/AAT) team.

Carole then single-handedly created a wealth of AAT resources in her community. She taught 4-H Club members how to train dogs. She became a certified AAT evaluator herself, and then certified her 4-H charges in AAT. From this group she created the AAA/AAT volunteer group, Walsh Winners 4-H Animal-Assisted Activities. She organized their outings and, with a gaggle of 4-Hers

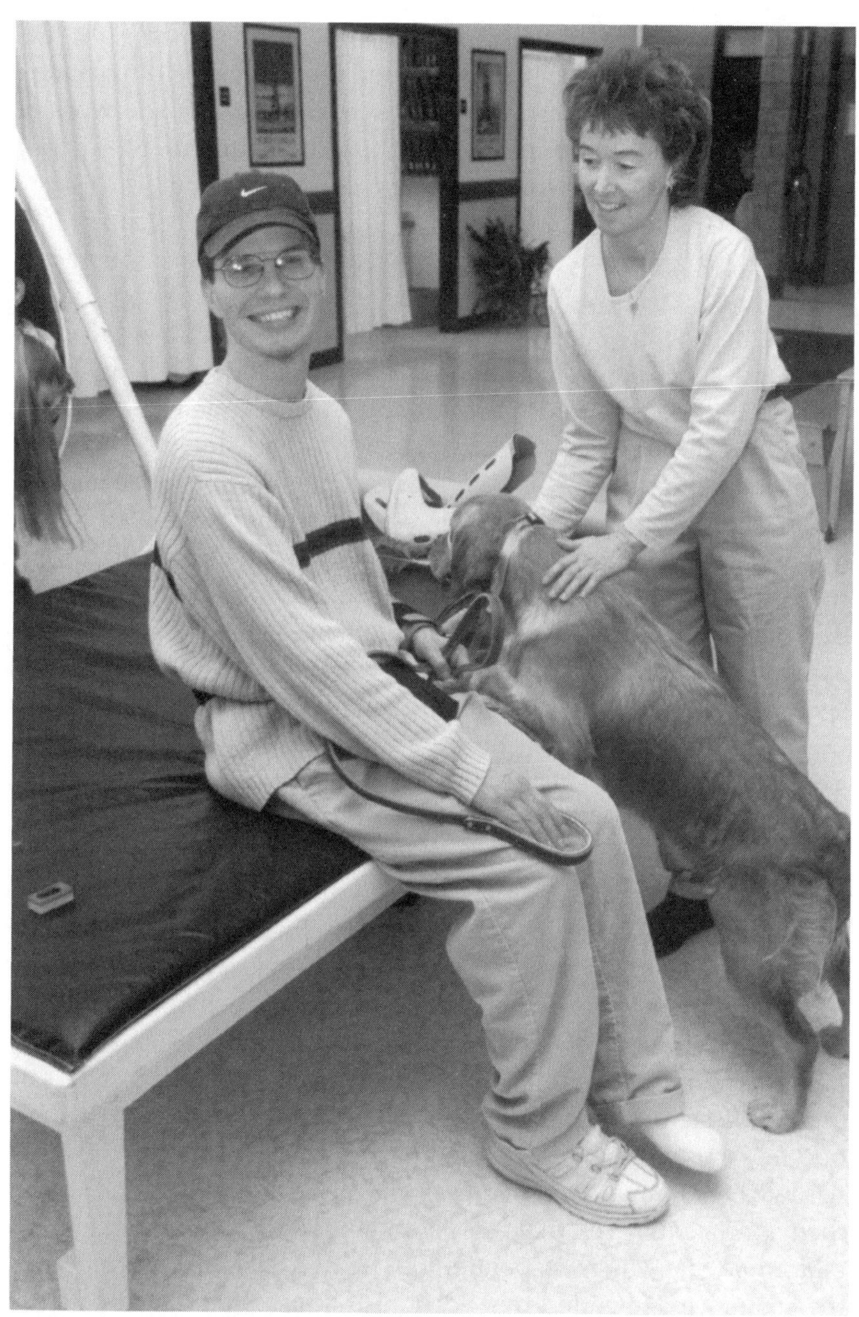

Lorn takes off his braces for physical therapy with Missie and Carole.
(Photo by Donald W. Smith)

in tow, visited people living in the local nursing home. At the public elementary school Carole created the "Paws Helping Hands to Read Program" for the extended school program; she worked one-on-one with a child in the special needs classroom. She did all this in her spare time, while running a child-care program.

Then Carole took Missie to volunteer in the occupational therapy program alongside Licensed Occupational Therapist Tracey Johnson, at the North Dakota Developmental Center. Tracey needed help with clients who were tired of the repetitive OT exercises and she wanted to concentrate on a particular client: Lorn. Having learned about Lorn's needs and goals, Carole went home to develop a list of tricks to teach Missie. Missie already knew the basic obedience tasks—sit, stay, down, come—but now she needed to, on command, touch the end of a stick with her nose, play "hide-n-seek" with a battery operated cow, play tug of war with a rope tug-toy, retrieve an earned treat from someone's hand when they said "take it," and follow some simple hand signals. Each of these actions was designed to help Lorn work on a particular motor, speech, or cognitive task, sometimes on all three at once.

On the day we visited Lorn he was already there, waiting for Missie. He had earned the opportunity to work with Missie by adhering to his work rules and by working on his personal goals. Lorn worked first on brushing Missie, a job that extends his arm across his body to strengthen and lengthen his normal range of motion. He worked primarily with his weaker left hand. Brushing Missie also strengthened the grasping muscles in Lorn's hands. Missie's job was to lay on her side and let him brush her. Next Lorn worked on "teaching" Missie, with verbal commands and accompanying hand signals: "sit," "down," "roll on your side," and "roll over." Missie already knows the four commands Lorn gives her, but the focus was not really on her learning. It was on Lorn remembering the commands, doing them in the correct order, and doing them sufficiently well that Missie understood. Lorn was able to do all four so that Missie responded but he needed help remembering the sequence. He ended this portion of the session with treats for Missie which, Lorn is reminded, he can let her have after he says,

"Missie, take it." Lorn had no problems enunciating this clearly and loudly enough so that Missie responded and he laughed like someone whose world is perfect.

Lorn had to think about many things while he was standing in front of Missie. He had to check his balance—he tends to slump to the left—he had to coordinate his hand and fingers to clearly communicate the hand signals, and he had to do them in the correct order. Missie could not, for example, have responded to the roll over command if she had not already been on her side. Having Missie present in Lorn's therapy gets him focused on her, not on the dreary repetitive therapy tasks. Her responses motivate him to continue. Missie is a happy and willing participant and Lorn obviously likes having her there (he wants to one day work with animals). In addition, each task accomplished something and made the whole more meaningful to Lorn. A half hour before this session, another therapist had been working with Lorn on the same kinds of tasks. In that session Lorn needed thirty-nine cues over the hour, more than one every two minutes, to stay on task. Now, with Missie, Lorn is fully engaged.

The normal forty-five-minute OT session stretched to ninety minutes and Lorn was still not ready to stop. The whole time he was focused and happy; he needed fewer than ten reminders during the hour and a half, did all the exercises willingly and for longer than required, and put in considerably more effort. His progress, Tracey says, is starting to extend into his work day. He pays attention longer and works harder at his job on the days that Missie comes to see him in therapy and he is better at controlling his impulses.

Lorn's desire is to help kids so they don't end up like him. He makes trips, accompanied by a staff person, to talk to kids at the local schools about his own drug abuse. Standing in front of a class, he struggles to make his words understandable. He is a potent reminder that drugs can destroy a person's most important organ, their brain. Although Lorn will never be the man he once was, the man he is now communicates clearly the mistakes he made by getting hooked on drugs. Working with Missie helps Lorn remember what he wants to say and articulate his message more clearly, so he can touch the hearts of adolescents who might think twice about chemical experimentation.

Chapter
Twenty-Two

Hoss

A friend is someone who knows the song in your heart and can sing it back to you when you have forgotten the words.
Unknown

The ferry ride was only part of the journey for those waiting at the dock to see where their loved ones perished, to acknowledge their profound loss, to say goodbye, and, hopefully, to realize that they were not alone. That was the intent of the police commissioner from the mayor's office of New York City when she sought out the comfort dogs that first day. She wanted them specifically and intended that there be one support escort for each family that would be traveling to Ground Zero for the first time. Ideally, each family would have someone who could offer the level of support needed. On one ferry that left for Manhattan that morning were Josiah and his German Shepherd therapy dog, Hoss.

Josiah and Hoss, along with Cindy Ehlers and Tikva (see chapter 2), were the first two therapy dog teams called to offer

Hoss at Ground Zero with an NYPD officer, September 2001. (Photo by Josiah Whitaker)

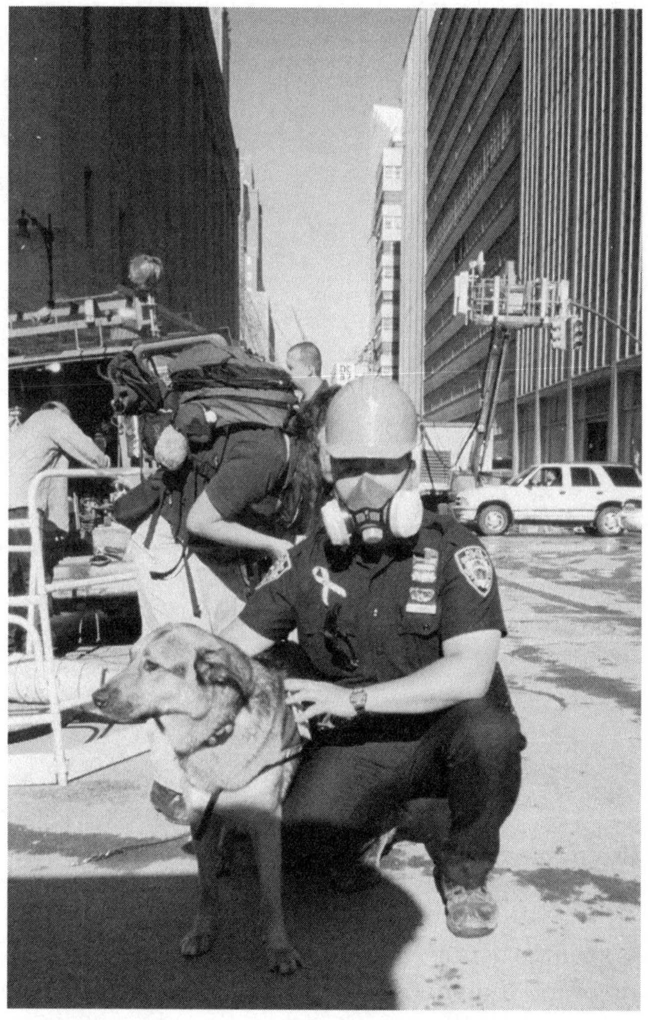

solace at Ground Zero. So great were the dogs' powers for soothing, that the awareness of their presence traveled faster each hour. Throughout the bleak days, firefighters, police officers, emergency workers, and others were grateful for the chance to touch the dogs, talk about their own families, remember their own animals which they had not seen for days, and, most profoundly and seemingly for the first time, to speak aloud the grief they felt and the trauma they had experienced. The therapy dogs were christened by the workers, "comfort dogs."

Karen Soyka, Red Cross Disaster Mental Health Volunteer, with Tikva and Hoss, September 2001. (Photo by Josiah Whitaker)

Since their arrival in New York on September 22, both comfort dogs and their owners had worked steadily, offering what consolation they could whenever the occasion allowed. For ten to twelve hours of each of Josiah and Hoss's first three days in Manhattan, the scenario of loss revealed and comfort given repeated again and again. At the end of the third day they were pulled away to work directly with the families of the victims.

They arrived early the next morning at the Family Assistance Center. Before three hours had passed, the police commissioner had learned of the comfort dogs and sought out Josiah. She wanted Josiah and Hoss's participation on a team she was organizing. Three

Emergency workers at Ground Zero with therapy dogs Tikva and Hoss. (Photo by Josiah Whitaker)

times a day, she said, a ferry would bring about fifty family members to Ground Zero. On each ferry ride about ten chaperones would provide needed support to the families during their journey. The escorts included members of the clergy, mental health workers, emergency medical technicians (EMTs), and the comfort dog teams. As the families arrived and waited to board the scheduled ferry, each member of the support team would seek out a family to whom they felt a natural bond, until all families had been accounted for and all families had a support person there to assist them.

Across the several trips he made, Hoss's role was seldom the same for any two families; his presence, though, was important to all of the families that he and Josiah assisted. Josiah has described what he and Hoss do as a dance; the choreography of the two is intuitive, with first one leading and then the other, taking their cues and reacting in a duet to the needs around them. On the ferries, as well as at Ground Zero, it was generally Hoss that led. For some families Hoss was a welcome diversion as they made their difficult emotional journey. For others he was a spiritual guide as they journeyed to the resting place of their loved ones and back again. Others saw Hoss in a more conventional role; he was a safe friend in whose fur they could bury their heads and cry, or who they could hold fast for as long as they needed without fear of judgment. To many of the families Hoss filled multiple roles. All of them saw Hoss as a welcome friend who had traveled from far away to offer them whatever he could in their time of need. At least one member of every family that Josiah met said as much to him as they traveled together, to and from Ground Zero.

One particular family, the mother, sister, daughter, and brother-in-law of a man lost in the attacks, formed a quick and natural bond with Josiah and his partner. They were simply attracted to Hoss. At the beginning of the twenty-minute ferry ride, they casually asked a few questions about Hoss. It appeared initially as if they wanted nothing more than a diversion from the terrible and imminent confrontation with their loss. The more they asked, though, the more they became convinced that Hoss embodied in some way their lost relation. Hoss was their brother, father, son, and was here to guide them, to help them pass over to the point of acceptance and to continue moving forward.

As they left the ferry Hoss no longer moved effortlessly by Josiah's side; he now paced a mourner's march at the side of the victim's mother, seemingly unaware of Josiah. He stayed with the family the entire time, never looking to Josiah even once to see if he was where he was meant to be. Arriving at the Ground Zero overlook, the family sobbed and embraced one another. Hoss was beside them, supporting them with his full attention. He remained within reach of the family, should someone want him, until he was again aboard the ferry. Hoss then found his place between the grieving women as they returned to the pier, giving them gentle kisses.

The family repeatedly expressed their extreme gratitude to Josiah and Hoss. They insisted they be allowed to drive Josiah to the airport, but Josiah, sensitive to their needs and respectful of their vulnerable state, felt obliged to refuse. So they sent letters to

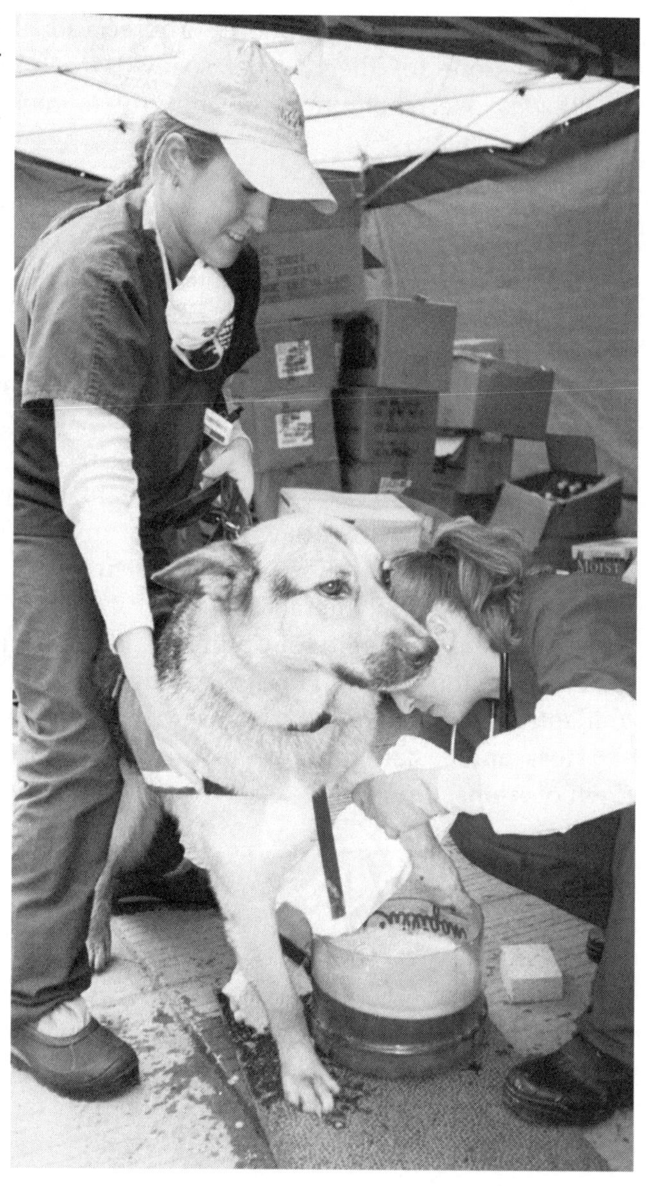

Volunteer workers bathe Hoss's feet at Ground Zero. (Photo by Josiah Whitaker)

Josiah, expressing their immense appreciation for what he and Hoss had done for them. What seemed woefully inadequate to Josiah was overwhelmingly meaningful to them, as was made clear in their words:

> I do want to thank you for all you and Hoss did for us the day we met. It made that day easier. My family will *never* [their emphasis] forget you or Hoss. It was not an accident that we met that day. My brother came to us through Hoss to tell us that everything will be OK. . . . I am grateful to whatever forces brought us together. Hoss was a great help at a very trying time in our lives.

In the days following the terror, the grief that blanketed Manhattan was much heavier than the weight of the Towers themselves. Josiah and Hoss knew they must involve themselves very directly in the pain of those who were suffering the most. "Yet," said Josiah, "in every interaction that I can recall, I recognized that others saw the possibility of optimism. Because of the presence of Hoss, a dog, the people [Hoss] worked with were better equipped to identify, acknowledge, and express their own humanity."

Hoss and Josiah continue to provide comfort in their home town of Aloha, Oregon.

Chapter

Twenty-Three

Zorro

At the Eastside Humane Society Zorro was deemed an unadoptable mutt. "Too rambunctious" was the reason given by each person who returned him. It looked like Zorro would become another animal statistic, until Megan Wolf, in the midst of some personal challenges, needed Zorro as much as he needed her. Their bond was instantaneous and complete.

Besides being a companion to Megan, Zorro had an uncanny knack for reassuring little boys and girls. This knack, complemented by successful behavior training, led Megan to believe that Zorro's gifts were ones that should be shared with children in need. Megan and Zorro tested for and received their therapy pet certification. Shortly thereafter they began working as part of the children's physical therapy team at Valley General Hospital.

Fast forward . . . three and a half years.

We are visiting Megan and Zorro today to see six-year-old Brendan Strunk, who is coming in for his weekly physical therapy. Due to a stroke suffered while in his mother's womb, Brendan

Lauren Adams, Brendan Strunk walking to Zorro, and Megan Wolf working on walking in physical therapy. (Photo by Karen A. Pomerinke)

faces many physical and emotional challenges. Even though physical therapy is difficult for him, Brendan looks forward to seeing his pal, Zorro, who makes therapy fun. However, our presence as observers has disrupted the routine that is crucial to Brendan. He balks as he sees us and begins crying loudly and resolutely:

"*No! I* don't want to! Not *anything*!"

Brendan's physical therapist skillfully diverts Brendan's attention to Zorro. Her posed questions are part of Brendan's routine and signal familiarity and comfort to him.

"Who is big and black and is waiting for you, Brendan?"

Brendan knows that Zorro is the big black dog happily anticipating him. The volume of his screaming quiets toward adamant statements.

"Do you know what his tail is doing, Brendan?"

Brendan is almost quiet, his eyes squeezed shut.

"Guess how he's holding his ears."

Brendan's eyes slowly open and shine with anticipation. He gazes past the therapist to the doorway leading to Zorro. His tears

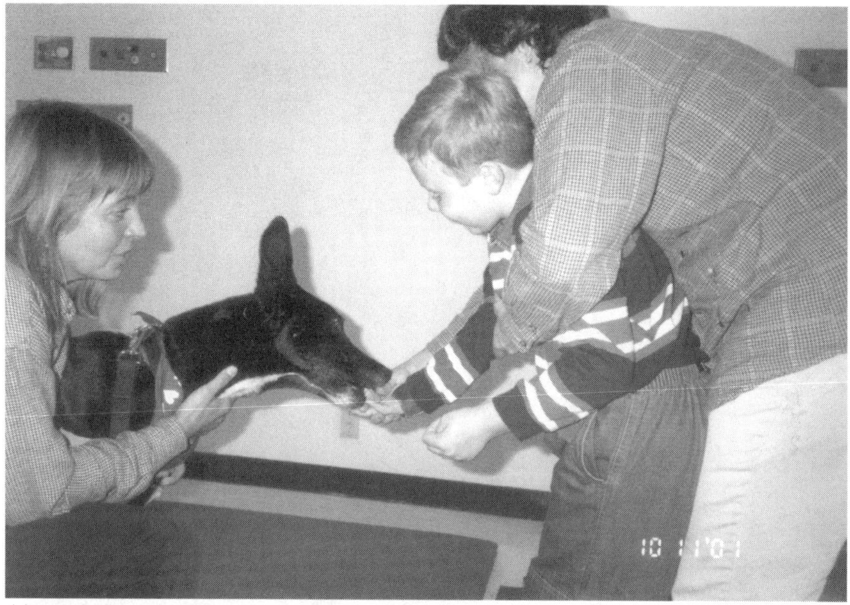

Megan, Brendan, and Lauren, feeding Zorro his treats.
(Photo by Karen A. Pomerinke)

forgotten, Brendan now smiles and leaves happily with his therapist, eager to see Zorro.

In the adjoining room, Brendan's physical therapy is orchestrated to seem like play to both Brendan and the casual observer. Brendan reaches down to grasp a ball and places it carefully in the hands of his therapist, who braces him from behind. She places the ball in a "dog-a-pault," a device designed to allow either humans or canines to launch the ball for catching. Brendan raises his foot up and stomps down on a wood slab, sending the ball flying to Zorro.

"Yeah Brendan!" comes the yell of his personal cheering squad. Their cheers are as much for the effort it takes Brendan to simply bend down and clutch the ball as they are for his launching the dog-a-pault without falling. Later, still more cheers are given for Brendan's efforts at walking on a balance beam toward Zorro.

Brendan's therapy culminates with everyone's favorite game, "hide-n-seek," with a twist. Brendan must walk backward to a spot of his choosing and hide a treat for Zorro. Zorro waits for him, eyes and ears covered, until Brendan yells the go-ahead: "O.K. Zorro! Find it!"

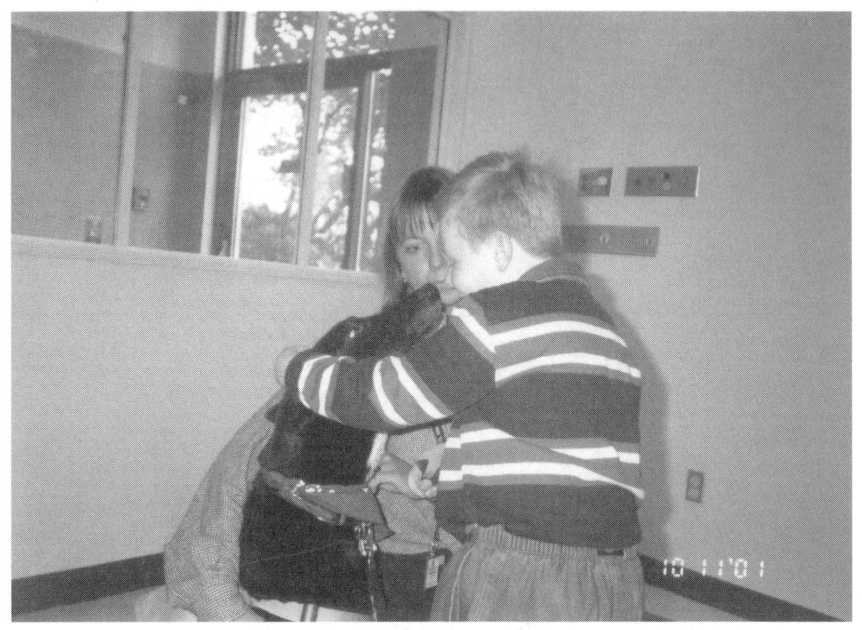

Brendan and Zorro share some mutual affection.
Megan is in background. (Photo by Karen A. Pomerinke)

Zorro's finding the treat signals the end of the session: more cheers for Brendan and for Zorro. As Brendan spends the last few minutes of his session touching and hugging Zorro, Zorro is only too delighted to reciprocate. Brendan laughs as Zorro's long pink tongue repeatedly kisses his cheek.

Over the last three and a half years, Zorro has been the catharsis for Brendan's mobility. Before Zorro came to therapy, Brendan could not walk. Afraid of falling, anxious about the pain, and worried that he couldn't do it, Brendan refused to participate in physical therapy. Zorro moved Brendan beyond his fears. He took his first steps because he was more motivated by the chance to pet Zorro and hold his leash than he was afraid of his potential failure and pain.

Brendan's progress continues in tiny increments. At one time, however, there was little hope for Brendan to ever walk. Brendan made and continues to make progress, a wonder to many. To others it is a miracle named Zorro.

Appendix I

Getting Started in Animal-Assisted Therapy

Animal-Assisted Therapy (AAT) is the process of including a well-trained and well-socialized animal as a therapeutic tool. Therapy animals are commonly used in medical, physical, and mental health settings as well as in rehabilitation. Initially they were incorporated as a final effort when the professionals were stumped. More and more often animals are now incorporated not only into healthcare, but also into law-enforcement, corporate plans for employee satisfaction, and into the last days of many who find comfort from the animals. Not only because of the remarkable success stories but because of an ever-growing body of research documenting the benefits of therapy animals, smart health management teams are acknowledging this value. They also see that therapy animals are exceptional when it comes to establishing an atmosphere of trust; to relaxing, calming, and motivating people; and to creating success when failure appears inevitable.

The dog, cat, or other pet that works as a therapy animal does not have to possess special talents or extraordinary gifts. Basic obedience in dogs is expected; reliability and nonaggressive behavior are musts. Shyness or general fear is not accepted in a therapy animal; these traits lend themselves to possible problems and generally do not make a person feel that the therapy pet wants to be near them. In short, a good therapy pet instills confidence in the client or patient. Their natural sensitivity toward people's nonverbal communications helps them to excel in therapeutic interactions.

Animal-Assisted Therapy is an amazingly successful technique. It produces results that frequently cannot be achieved by any other method. Therapy pets repeatedly show that animals can "jump start" therapy when the therapists, clients, or patients are avoidant, have become stuck in their therapies, or have little to no hope of progress. The benefits of therapy pets are legend in hospitals and nursing homes, the two places in which such pets were first introduced. The stories are of children who learned to walk because a dog was there to walk with them (physical therapy) or learned to dress themselves because they acquired the necessary fine motor skills by grooming a therapy cat (occupational therapy). The benefits occur across a wide spectrum of people with severe and chronic depression, who begin to talk, to approach others, to have hope because a therapy pet lay its head in their laps and "understood" (mental health). People suffering from disasters and those helping them cope with the stress of a devastating loss find a connection, a way of releasing their immense grief because a therapy pet is present as they struggle to carry on (disaster and crisis interventions). People struggling with chronic alcoholism find comfort and understanding when a therapy animal is present in their therapy group (chemical dependency treatment groups). Juveniles confined in treatment centers with little regard for the needs of others learn they can empathize and develop positive, caring relationships because a therapeutic horse or dog was part of their treatment (juvenile treatment centers).

Who would have thought it? Who could have guessed that therapeutic horse riding could improve not only someone's motor skills but their speech (verbal apraxia) and their ability to focus attention (attention deficit hyperactivity disorder)? Those with

HIV/AIDS have decreased stress and thus better immune functioning because an animal is part of their life. And people in comas have been nursed back to consciousness because therapy animals were part of their medical treatment.

For thousands of years people have developed bonds with animals and now that bond is being extended beyond the owner-pet relationship with positive effect. Pets have always been those we turn to when we cannot speak to anyone else. Anyone who has felt this deep and steadfast bond with an animal would not be surprised that therapy pets forge these same emotional connections with people other than their owners. This is the secret behind every one of the stories in this volume. It is the reason that therapy pets are being incorporated in all of the above settings and many others.

Where do therapy pets come from? They are family pets or farm animals (for example, llamas) that have been tested and certified by an organization to meet specific and stringent behavioral, socialization, and health standards. The animal, under the guidance of the person who accompanies it to the testing, is predictable, steady in the face of unexpected events, and friendly. The person can direct the behavior of the animal in a knowledgeable and safe fashion. The animal is clean and well groomed, free of illness and infection; the handler maintains health standards for both the animal and the client. Although the certification process varies from animal to animal and even across different therapy organizations, the golden rule remains constant: the animal-human team must help to inspire confidence in the client.

In general, certification involves passing both a knowledge test and an animal-handler performance test. The knowledge test is generally geared toward assuring that the handler knows basic healthcare regulations, transmittable diseases from humans to the pet, as well as guidelines for visiting. Most organizations have written material available for therapy pet handlers to study. Many evaluators have courses and classes available for those who wish to prepare for testing. There are age limitations for pets. Dogs, cats, horses, and other animals are expected to be at least one year old.

There are hundreds of therapy pet programs available if you

and your pet plan to begin therapy work, and we encourage you to set up and standardize your own local club. You can get started by contacting one of the certifying organizations or local groups listed in appendix 3. National certifying organizations have evaluators located throughout the United States and many other countries around the world. The contact information for the evaluator closest to you can be provided by the national certifying organization; they can also tell you what to prepare for and how to go about it. Make the choice to make a difference with your pet today.

Appendix

II

Standards and Terms Associated with Therapy Pets

The inclusion of trained and certified animals in combination with volunteers or licensed professionals has evolved into a field with a solid reputation, but whose name is still under discussion. As a result of this ongoing discussion there is some natural confusion in the general public about what makes an animal beneficial to a person's health and what you call them. As the title of this book implies, we prefer "Therapy Pets." It is simple and straightforward. Other common terms and an explanation of what makes an animal beneficial to a person's health are described below along with the generally accepted principles and criteria used in the United States and abroad. Most importantly, we stand behind a nationally accepted standard with periodic reassessments of the animal-human teams and specifiers that could limit a team to certain kinds of environments and/or populations if appropriate. Although there is not presently one nationally accepted set of standards, those who believe in the value of predictable animals helping heal humans are working toward that standardization.

GENERALLY ACCEPTED STANDARDS:

The animal-human team has been certified by a nationally recognized organization (either local or national) to have met the minimum behavior and socialization criteria for intervening with the general public. Recertification of the team is required at regular intervals. Certification can be limited to certain types of environments and populations.

For dogs the minimum behavior and socialization criteria is based on the American Kennel Club (AKC) Canine Good Citizen (CGC) test. For other animals, the criteria is based on the animal being under the control of the handler at all times and not subjected to situations that are overly stressful to the animal.

The basis for the socialization is graduated, multiple contacts between the animal to be certified and people and other animals in a way that ensures safety and pleasure in these same contacts. Ideally, these interactions begin as soon as the animal is old enough to tolerate the contacts safely and well before the animal reaches maturity.

RELATED TERMS:

Service animals—The Americans with Disabilities Act (ADA) defines a service animal as "any guide dog, signal dog, or other animal individually trained to provide assistance to an individual with a disability." If they meet this definition, animals are considered service animals under the ADA regardless of whether they have been licensed or certified by a state or local government. Service animals perform some of the functions and tasks that the individual with a disability cannot perform for him or herself. "Seeing Eye Dogs" are one type of service animal, used by some individuals who are blind. Less common, but just as valid, are service dogs to help people with debilitating emotional disorders, such as dogs who accompany people with agoraphobia into the community. Service animals are trained to eventually be owned by and assist one person with disabilities. Certified therapy pets are not automatically service animals. These are two separate kinds of certification.

Assistance animals—An assistance animal is a newer term being proposed to replace the term "service animal." It is similar to a service animal but instead of limiting the animal to assisting one person with a disability, an assistance animal works either with a specific disabled person or a group of people with disabilities, under the guidance of a trainer or owner. The animal's training is similar to that given service animals. As with service animals, certified therapy pets are not automatically assistance animals. These are two separate kinds of certification

Therapeutic Riding—A general category of assisted riding designed to improved physical, psychological, and social functioning. Therapeutic riding normalizes muscle tone, increases flexibility, improves coordination, balance and strength, circulation, respiration, and posture. Other improvements are common in self-esteem, self-confidence, and self-control, in spatial orientation, fine and gross motor control, in social interaction, and mental relaxation

Hippotherapy—Hippotherapy is therapeutic riding that requires face-to-face supervision by a state licensed or certified physical or occupational therapist. It combines horseback riding with therapeutic exercises to improve the rider's posture, coordination and balance, to normalize muscle tone and strengthen muscle groups. The benefits of hippotherapy are rooted in the way a horse's gait mimics the rhythm of normal human walking. The horse's movements provide feedback to the rider's neurological system for the pattern of normal movement. That feedback, in turn, influences the rider's posture, muscle tone, balance, motor functioning, and sensory processing.

OTHER COMMONLY ACCEPTED NAMES FOR THERAPY PETS:

Pet Therapy

Pet Assisted Activity and Therapy (PAA & PAT)

Pet Facilitated Activity and Therapy (PFA & PFT)

Animal-Assisted Activity and Therapy (AAA & AAT)

Animal-Facilitated Activity and Therapy (AFA & AFT)

There are pros and cons to each of the above terms and their use. My own sense is that the use of the word "pets" is easily understood by almost everyone and adequately defines what the animal is. The word "pet" connotes an animal friendly to humans. Those who maintain that the word "pet" is self-limiting and does not include a broad enough spectrum of the animal kingdom are, to my way of thinking, putting up false arguments. Any animal that has learned from an early age that humans do not act toward them in malice (indeed, that *their person* will protect them) and therefore enjoys being in the presence of humans, is a pet. It can become a therapy pet if it also learns to follow the rules of expected behavior around humans and to enjoy the presence and touch of other people, in addition to that of their owner.

What about the difference between *assisted* and *facilitated*? *Assisted* suggests greater involvement in the process. *Facilitated* is more passive. The definitions are not far from each other, though, and to try to differentiate them seems like haggling.

The terms *activity* and *therapy*, have better discrimination and are less debated as to their general intention in combination with either *pet* or *animal*. A *pet-assisted activity* is any therapeutic animal-human contact, regardless of whether it results in a positive or health-engendering change, that is done under the guidance of a lay person (that is, a nonprofessional, unlicensed person). Most people involved in *pet-assisted activities* are volunteers. An example is an animal-handler team visiting the elderly in a nursing home; in this instance the pet-handler team has no one guiding the visit and there is no goal to be reached.

Pet-assisted therapy, on the other hand, may very well look the same, but is designed to reach a predetermined goal and is carried out under the auspices of a state licensed or state certified professional. In the previous case, the visit would be considered therapy if the elderly person had a goal (for example, to increase normal mood for five out of six weeks) and a licensed person supervising the time spent working toward that goal.

The difference between *therapy* and *activity* looks like nitpicking until one considers the question, who will pay for these services? If there is going to be a payee, then the service provider needs to show that the pet-assisted therapy is (1) needed and (2) beneficial. In some organizations, the administration may also want to show that the service is capable of making a profit. For these reasons there is a need for goals and monitoring.

The Delta Society, a national certifying organization, has chosen to use the term "animal-assisted activities" (AAA) for those actions that enhance quality of life for any person and are not goal directed. They use the term "animal-assisted therapy" (AAT) for "a goal-directed intervention in which an animal meeting specific criteria is an integral part of the treatment process." Many people use these terms, many more advocate them. I have found that I use these terms with people who are "in the biz." Otherwise, the terms are bulky and meaningless with most other people, and with the uninitiated I prefer to use the term "pet therapy."

Appendix

III

National and State Listings of Therapy Pet Organizations

NATIONAL ORGANIZATIONS

Local therapy pet organizations are likely to be greatly expanded upon by the time of publication of this book. Nonetheless, the lists below provide a starting point for you. Other starting points include contacting local kennel clubs and hospital volunteer departments that may have therapy pet teams already. If you're looking for the most up-to-date information, check online listings at www.therapypets. com.

North American Riding for the Handicapped Association (NARHA)
P.O. Box 33150
Denver, CO 80233
www.narha.org

Therapy Dogs International, Inc. (TDI)
88 Bartley Square
Flanders, NJ 07836
www.tdi-dog.org

Delta Society
289 Perimeter Road East
Renton, WA 98055-1329
www.deltasociety.org

Therapy Dogs, Incoporated.
P.O. Box 5868
Cheyenne, WY 82003
www.therapydogs.com

STATE ORGANIZATIONS

ALABAMA

Hand in Paw
Village East Shopping Center
5342 Oporto Madrid Boulevard South
Birmingham, AL 35210
www.handinpaw.org/

Pets and People: Companions in Therapy and Service
P.O. Box 40143
Mobile, AL 36640
www.petsandpeople.org

Special Equestrians
900 Woodward Drive
Indian Springs, AL 35124
http://home.hiwaay.net/~scottj2/spclequest2.html

ARKANSAS

Pawsitive Connection

1207 East Elm Street
Fayetteville, AR 72703
www.geocities.com/pawsitiveconnection/

ARIZONA

Companion Animal Association of Arizona
P.O. Box 5006
Scottsdale, AZ 85251-5006
www.caaainc.org

Gabriels's Angels, Inc.
4855 E. Warner Road
Suite 24, PMB 102
Phoenix, AZ 85044
www.petshelpingkids.com

Pets on Wheels of Scottsdale, Inc.
7375 East Second Street
Scottsdale, AZ 85251
www.petsonwheelsscottsdale.com

Therapeutic Riding of Tucson (TROT)
P.O. Box 30584
Tucson, AZ 85751
www.horseweb.com/client/trot/index.htm

Tucson Area Pet Partners
10567 N. Camino Rosas Nuevas
Tucson, AZ 85737
www.volunteersolutions.org/agency/one_178848.html

CALIFORNIA

CC/SPCA Pet Facilitated Therapy
103 S. Hughes
Fresno, CA 93706-1207
www.unitedpaws.com/ccspca/therapy.html

"Create-a-Smile" Animal-Assisted Therapy Team
237 Hill Street
Santa Monica, CA 90405
www.create-a-smile.org

Foundation for Pet-Provided Therapy
P.O. Box 6308
Oceanside, CA 92058
www.loveonaleash.org

Friendship Foundation
P.O. Box 6525
Albany, CA 94706
www.friendship-foundation.org

Furry Friends Pet-Assisted Therapy Services
P.O. Box 5099
San Jose, CA 95150
www.furryfriends.org

Lend a Heart—Lend a Hand Animal-Assisted Therapy, Inc.
P.O. Box 60617
Sacramento, CA 95860
www.lendaheart.org

Loving Animals Providing Smiles (LAPS)
336 Twin Oaks Drive
Napa, CA 94558
E-mail: LAPS_AAT@msn.com

Ohlone Humane Society "Hug a Pet" Therapy Program
PMB #108
39120 Argonaut Way
Fremont, CA 94538
www.ohlonehumanesociety.org/hugapet.htm

Pam Lemke-Wright
Santa Cruz, CA
www.placenet.net/therapydogs.html

Paws'itiveTeams Therapy Dog Prep School
P.O. Box 27018
San Diego, CA 92198
www.pawsteams.org/therapy.htm

San Francisco SPCA Animal-Assisted Therapy Program
2500 16th Street
San Francisco, CA 94103-4213
www.sfspca.org/aat/index.shtml

SPCA Los Angeles Animal-Assisted Therapy Program
5026 West Jefferson Boulevard
Los Angeles, CA 90016
www.spcala.com/pages/aatherapy.htm

TherapyPets
P.O. Box 32288
Oakland, CA 94604-3588
www.therapypets.org

COLORADO

Cadence Center for Therapeutic Riding
P.O. Box 9009
Durango, CO 81301
www.creativelinks.com/cadence

Colorado Boys Ranch
P.O. Box 681
La Junta, CO 81050
www.coloradoboysranch.org/cbr/animals.html

Denver Pet Partners
P.O. Box 270113
Littleton, CO 81207
www.denverpetpartners.org

Mita Sunke Equine Learning Center
1990 West 150th Avenue
Broomfield, CO 80020
www.mitasunke.org/

Table Mountain Animal Center
Pet Therapy Program
4105 Youngfield Service Road
Golden, CO 80401
www.tablemountainanimals.org/volunteers.html

CONNECTICUT

Tails of Joy Therapy Dog Program Connecticut
c/o Susan Gagnon
233 Willington Hill Road
Willington, CT 06279
http://www.tailsofjoy.org/index.htm

FLORIDA

Freedom Ride, Inc.
P.O. Box 3741
Winter Park, FL 32790-3741
www.freedomride.com

Gulf Coast Chapter Delta Society Pet Partners
10501 FG-CU Boulevard South
Ft. Myers, FL 33965-6565
www.fgcu.edu/cfpa/pettherapy/index.html

Hug A Pet Therapy Group
Pasco County, FL
www.hugapet.freeservers.com

Sarasota Manatee Association for Riding Therapy (SMART)
P.O. Box 9566
Bradenton, FL 34206-9566
www.smartriders.org

GEORGIA

Dreamworkers
4704 Brownsville Road
Powder Springs, GA 30127
www.dogsaver.org/dreamworkers/

Happy Tails Pet Therapy
P.O. Box 767961
Roswell, GA 30076
www.happytailspets.org

Therapeutic Dogs of Georgia
E-mail: therapydogsofgeorgia@msn.com

IOWA

Miracles in Motion
P.O. Box 14
Cedar Rapids, IA 52406-0014
www.miraclesinmotion.net

ILLINOIS

Chenny Troupe, Inc.
1504 North Wells Street
Chicago, IL 60610
www.chennytroupe.org

Cowboy Dreams
241 Otis Road
Barrington Hills, IL 60010
and Kickapoo Farms
31W952 Penny Road
Dundee, IL 60118
www.cowboydreams.com

The Lincolnshire Animal Hospital Pet Visitors Group
Lincolnshire Animal Hospital
420 Half Day Rd.
Lincolnshire, IL 60069
Phone: (847) 634-9250
No Web site
E-mail: FEPMGP@aol.com

Pegasus Special Riders
Flagg and Carthage Roads
Oregon, IL
www.pegasusspecialriders.org

INDIANA

People & Pets Together
214 E. Maple Street
Jeffersonville, IN 47131
www.therapydog.org

KANSAS

Serenata Farms School of Equestrian Arts—Horses Helping Humans
1895 East 56 Road at Big Springs
Lecompton, KS 66050
www.serenata.org

KENTUCKY

Exceptional Equitation
Spruce Point Farm
2107 Massie School Road
La Grange, KY 40031
www.exceptionalequitation.org

Wags Pet Therapy of Kentucky, Inc.
P.O. Box 91436
Louisville, KY 40291-1436
www.kywags.org

LOUISIANA

Visiting Pet Program
5831 South Johnson Street
New Orleans, LA 70125
www.visitingpetprogram.org

MASSACHUSETTS

Dog B.O.N.E.S.
17 Bradshaw Street
Medford, MA 02155
www.therapydog.info

White Oak Farm
411 North Street
Jefferson, MA 01522
www.whiteoakhorsefarm.com

MARYLAND

Back to Fitness Therapeutic Riding
Eldersburg, MD
www.bcpl.net/~gharris/ther.html

Great Strides Therapeutic Riding
P.O. Box 451
Damascus, MD 20872
www.greatstrides.org

MAINE

Equest Therapeutic Riding Center
P.O. Box 935
Kennebunk ME 04043
www.equestmaine.org

Flying Changes Center for Therapeutic Riding
Route 201
Topsham, ME 04086
www.flyingchanges.org

MICHIGAN

Children and Horses United in Movement (CHUM)
P.O. Box 14
Mason, MI 48854
www.chumtherapy.org

Offering Alternative Therapy with Smiles (OATS)
3090 Weidemann Drive
Clarkston, MI 48348
www.oatshrh.org

MINNESOTA

Bark Avenue on Parade
P.O. Box 24071
Minneapolis, MN 55424
www.barkavenue.org

Helping Paws of Minnesota, Inc. Service Dogs
P.O. Box 634
Hopkins, MN 55343-0634
www.helpingpaws.org/AAAProgram.htm

Mounted Eagles Therapeutic Riding Program
3800 Blackbear Road
Brainerd, MN 56401
www.mountedeagles.org

Pals on Paws
1405 55th Street, NE
Rogers, MN
www.geocities.com/Heartland/Meadows/1442/Pals.htm

MISSOURI

Magic Moments Riding Therapy
394 County Lane 125
Diamond, MO 64840
http://www.geocities.com/heartland/ridge/9220/

Pet Therapy of the Ozarks, Inc.
P.O. Box 9462
Springfield, MO 65801
www.pettherapyozarks.homestead.com

Support Dogs, Inc.
3958 Union Road
St. Louis, MO 63125
http://members.aol.com/maxidog1/therapy.htm

NEBRASKA

Heartland Equine Therapeutic Riding Academy
29430 Ida Street
Valley, NE 68064
www.HETRA.org

Paws for Friendship, Inc.
P.O. Box 12243
Omaha, NE 68152
http://pawsforfriendshipinc.org

NEW HAMPSHIRE

Dog Logic Therapy Dog Team Training Program
3020 Brown Avenue #10
Manchester, NH 03103
http://www.doglogic.com/therapymain.htm

NEW JERSEY

Unicorn Handicapped Riding Association
Looking Glass Farm
40 Cooper-Tomlinson Road
Medford, NJ 08055
www.users.voicenet.com/~unicorn

NEW MEXICO

Cloud Dancers of the Southwest
Equestrian Therapy and Recreation for the Disabled
Albuquerque, NM
http://home.ATT.NET/~C-DSW

Southwest Canine Corps of Volunteers
Albuquerque, NM
www.nmia.com/~dmiller

NEW YORK

Astride, Inc.
P.O. Box 5241
Syracuse, NY 13220

Long Island Riding for the Handicapped
P.O. Box 352
Glen Head, NY 11545
www.lirha.com

Pal-O-Mine Equestrian
33 Lloyd Harbor Road
Huntington, NY 11743
www.pal-o-mine.org

Winslow Therapeutic Riding Unlimited
340A South Route 94
Warwick, NY 10990
http://winslow.org/winslow

OHIO

Doggie Brigade
Akron Children's Hospital
One Perkins Square
Akron, OH 44308
www.akronchildrens.org/job-volunteer/doggie.html

Equine Assisted Therapy
7908 Myers Road
Centerburg, OH 43011
www.equineassistedtherapy.org

Miami Valley Pet Therapy Association
3614 Knollwood Drive
Beavercreek, OH 45432
www.mupta.org

OKLAHOMA

Paws for Friendship
Oklahoma City, OK
www.geocities.com/pawstherapydogs/OKCtherapydogs. html

OREGON

Adaptive Riding Institute
P.O. Box 280
Scotts Mills, OR 97375
www.open.org/~horses88

Equitopia
1885 NW 9th Street
Corvallis, OR 97330
www.equitopia.peak.org

Project Pooch
Oregon Youth Authority MacLaren School
2630 North Pacific Highway
Woodburn, OR 97071
www.pooch.org/

RideAble
85873 Lorane Highway
Eugene, OR 97405
http://users.rio.com/aaj/index.html

PENNSYLVANIA

Animal Friends Pet Therapy Program
2643 Penn Avenue
Pittsburgh, PA 15222
www.animal-friends.org/site/petassist.jsp

The Capital Area Therapeutic Riding Association (CATRA)
P.O. Box 339
Grantville, PA 17028
www.catra.net

SOUTH CAROLINA

Charleston Counseling & Support Services
P.O. Box 30082
Charleston, SC 29417
www.geocities.com/animalassistedtherapy

SCDogs Therapy Group
Clemson University
Clemson, SC
www.scdogs.org

TENNESSEE

Shangri-La Therapeutic Academy of Riding (STAR)
P.O. Box 22453
Knoxville, TN 37933
www.korrnet.org/staride

TEXAS

All Star Equestrian Center
6601 FM 2738
Mansfield, TX 76028
www.geocities.com/allstarfound

Faithful Friends Animal-Assisted Therapy (AAT) Ministry
207 Briarwood Court
League City, TX 77573
www.faithfulfriendsaat.org

The Healing Hounds, In association with The Dog Training Club of Dallas County, Inc.
604 Crestside
Duncanville, TX
(972) 780-7698
No Web site as of this printing. Contact DTCDC Web site at www.dogpresident@dallasdogtraining.org

Ride On Center for Kids (ROCK)
P.O. Box 2422
Georgetown, TX 78628
www.rockride.org

Riding Unlimited
9168 T.N. Skiles Road
Ponder, TX 76259
www.ridingunlimited.org/

Self-Improvement through Riding Education (SIRE)
Route 2, Box 56
Hockley, TX 77447
http://members.aol.com/sireinc/index.html

SPCA of Texas and Dallas County Youth Village
Dog Obedience Training Program
362 South Industrial Boulevard
Dallas, TX 75207

UTAH

Intermountain Therapy Animals
1555 East Stratford Avenue
Suite 400
Salt Lake City, UT 84106
www.therapyanimals.org

Utah Animal-Assisted Therapy Association
P.O. Box 18771
Salt Lake City, UT 84118-8771
www.aros.net/~uaata/

VERMONT

Therapy Dogs of Vermont (TDV)
7828 Vermont Route 105
East Charleston, VT 05833
www.therapydogs.org

VIRGINIA

Animal-Assisted Crisis Response Association
c/o Lois C. Hardy
5314 Sunrise Shore
Chincoteague, VA 23336
www.aacra.org/index.html

A Leg Up Therapeutic Riding Center for the Handicapped
P.O. Box 1257
Abingdon, VA 24212-1257

Paws for Health
The Medical College of Virginia
Richmond, Virginia
http://views.vcu.edu/paws/

The Shiloh Project
12210 Fairfax Town Center
PMB #902
Fairfax, VA 22033
www.shilohproject.org/

WASHINGTON

Reading with Rover
8902 222nd Street, S.E.
Woodinville, WA 98072
www.readingwithrover.com

Sirius Healing Animal-Assisted Activities and Therapy Provider and Trainer
12046 12th Avenue NE
Seattle, WA 98125-5014
www.siriushealing.com

WISCONSIN

Stable Hands Therapeutic Riding Program for the Disabled
3501 Swan Avenue
Wausau, WI 54401
http://webpages.charter.net/stablehands

CANADA

The Lanark County Therapeutic Riding Program
103 Judson Street
Carleton Place
Ontario K7C 2S5
Canada
www.therapeuticriding.ca

Little Bits Therapeutic Riding Association for Persons with Disabilities
P.O. Box 7426
Edmonton, Alberta
Canada T5E 6K1

Pacific Animal Therapy Society (PATS)
9412 Laurie's Lane
Sidney, British Columbia
Canada V8L 4L2
www.island.net/~patspets

Pacific Riding for the Disabled Association
1088 208th Street
Langley, British Columbia
Canada V2Z 1T4
www.prda.ca

The Pet Therapy Society of Northern Alberta
330-9768 170 Street
Edmonton, Alberta
Canada T5T 5L4
http://paws.shopalberta.com

UNITED KINGDOM-ENGLAND

Donkey Facilitated Therapy
The Donkey Sanctuary
Sidmouth, Devon EX10 0NU
United Kingdom
www.thedonkeysanctuary.org.uk/ds/aat.htm

Appendix

IV

Administrative and Professional Opinions of Pet Therapy

Susie Bailey, MFA, MSW
Assistant Professor, Department of Speech Communication Theatre and Dance
Kansas State University, Manhattan, Kansas
President, National Association of Drama Therapy

> Kodiak has added a new dimension to the work of the Barrier-Free Theatre company this year. Developing expressive communication skills is one of the major goals of [the people we work with]. In learning how to communicate with Kodiak, they had to make clear signals and vocal commands. Whenever they did, [Kodiak] would respond to the command. Their sensitivity to Kodiak spilled out into their interactions with each other.

Diane Cremering, Recreation Therapy
St. Therese Medical Center
Waukegan, Illinois

Pet Therapy is awesome! I have been using the services provided by Marilyn Putz and her dog, Tegan, for approximately six years. Tegan has visited several units over the years (examples: physical rehab., skilled nursing, pediatrics, oncology and psychiatry). The response to Tegan by stroke and amputee patients is remarkable to see. Stroke patients who are aphasic suddenly speak in complete sentences. Amputee patients seem to get an emotional lift from seeing Tegan function with his own amputation. Our psychiatric patients who suffer from depression and are frequently socially isolative often display a change in affect while visiting with Tegan. Depressed patients often will become more social and report feeling better after participating in Pet Therapy.

Karen Stays, Volunteer Coordinator
Swedish Hospital
Seattle, Washington

It's a real plus for our patients. Studies document [that] patients can go longer without pain medication with a visit from a therapy dog. It's a program we certainly would not want to be without. Lots of times patients come from out of town; they will call ahead and arrange a visit. It's a great program.

Scott Berman, M.D., Psychiatrist
Bethlehem, Pennsylvania

Patients who have been treated by both Kathy and me have all felt that they benefited from animal-assisted therapy. Brutus and Kathy have done some amazing work. When I was hospitalized for plasma exchange to treat a rare neurological disease, Brutus came to visit. I had a central venous catheter surgically implanted (a central line that lets large amounts of fluid be taken in and out) and I was being hooked up to a plasma exchange machine that washes the blood to remove certain antibodies. It was tiresome and a high anxiety time of my life. I really relaxed with the chance to sit and talk and play with Brutus. I have now seen numerous good studies in the medical literature supporting the effectiveness of animal-assisted therapy.

Karen Soyka M.S., LPC, CCDC
Ad Hoc Professor, University of Akron
Doctoral Student in Counseling Education and Supervision
Akron, Ohio

As a Licensed Professional Counselor and a member of the American Counseling Association and Northeast Ohio Counseling Association I wanted to take a moment and thank the therapy dogs and trainers who assisted me in NYC at Ground Zero. The work there was so horrific and the rescue workers were suffering such profound grief that on my own the connections with the men were difficult, which was where the beauty and love of the therapy dogs was imperative. The healing they brought to the broken-hearted was phenomenal and I will forever be grateful to Cindy Ehler, Tikva, Josiah, and Hoss for their help.

Anjie Lindblom, Special Programs/Operations Assistant
Ronald McDonald House of Dallas, Texas

The Healing Hounds therapy dogs have been coming to visit the families of the Ronald McDonald House of Dallas every month for the past three years. The therapy dogs provide a level of therapy to the children of the house that cannot be duplicated. The children wait anxiously for the dogs and they ask repeatedly, "When will the dogs be coming?" Every holiday, as well as birthdays, brings the dogs in wonderful costumes. The joy they bring to the children is overwhelming and they bring smiles to children's faces during a stressful time of their lives. The visually and hearing-impaired children respond to the therapy dogs in ways that cannot be explained. The dogs communicate with these children like no one else can. The therapy dogs bring a remarkable service to our Ronald McDonald House and to the families.

Trish Janus, Supervisor
Physical Disabilities Program
Special Education
Montgomery County Public Schools
Bethesda, Maryland

> I have been most impressed with how motivated the kids are to work with Marley. Animals are nonjudgmental and both kids and adults naturally respond to that. While the animal-assisted therapy is not an end in itself, it can be an extremely effective enabler in achieving functional outcomes. The benefits are not just physical but social and emotional as well.